Martin H. Wolfson

STEEL

STEEL

UPHEAVAL IN A BASIC INDUSTRY

Donald F. Barnett
and
Louis Schorsch

Ballinger Publishing Company • Cambridge, Massachusetts
A subsidiary of Harper & Row, Publishers, Inc.

International Standard Book Number: 0-88410-397-8

Library of Congress Catalog Card Number: 83-10012

Printed in the United States of America

Library of Congress Cataloging in Publication Data

Barnett, Donald F.
Steel: upheaval in a basic industry.

Includes bibliographical references and index.
1. Steel industry and trade–United States.
I. Schorsch, Louis. II. Title.
HD9515.B36 1983 338.4'7669142'093 83-10012
ISBN 0-88410-397-8

CONTENTS

LIST OF FIGURES

LIST OF TABLES

ACKNOWLEDGMENTS

The authors would like to express their appreciation to the many individuals without whose contributions this book could not have been written. In particular, we are grateful for the insights of colleagues within the steel industry who have lived through the changes described in this book and who are now grappling with the question of the industry's future. While some may disagree with our conclusions, we trust that all will agree that the problem deserves to be rethought in a fundamental way.

We would like to thank several individuals who reviewed early drafts of the book; Donald Belch, Robert Crandall, Karlis Kirsis, and Laurence Schorsch deserve special mention. Judith Kaplan prepared the charts and figures, cheerfully coping with a stream of minor revisions. Phyllis Varipapa provided valuable support on several fronts, as did our editors, particularly Steven Cramer and Julie Doohan. This list is representative rather than exhaustive; omissions should not be interpreted as lack of appreciation. Finally, we owe a great deal to those whose assistance—and endurance—have been most crucial: Elizabeth Barnett and Kathy Berger.

In the broadest sense, our greatest debts are to our parents. In recognition of that fact, this book is dedicated to Louis J. Schorsch and to the memory of Amey B. Barnett.

I A DIS-INTEGRATING INDUSTRY

1 INTRODUCTION

The American economy—and indeed the whole industrial world—is now in the throes of major structural changes, potentially representing a third industrial revolution. The first industrial revolution, circa 1750–1820, involved replacing brute force with machines; its most salient features were the development of steam power, spectacular productivity increases in the textile industry, and the rise of the factory system. The second, circa 1870–1920, involved heavy mechanization, continuous assembly, and the development of new sources of power (electricity) and new forms of transportation (the automobile). We are now in the midst of a third industrial revolution, dominated by the spectacular development and application of advanced electronics (data processing, robotics, etc.). Already there is evidence of fundamental and permanent changes in the types, location, and relative importance of industries. Major changes in the skills required of labor and the location of the workplace—e.g., from urban offices to regional work centers or homes—are also occurring. Such changes portend dramatic alterations in the institutions and structure of society, the first signs of which are already upon us.

Few industries will see greater changes in the next decade than will the steel industry. The forces of change have been actively at work in this industry since the 1950s, while many other basic industries were spared until the late 1960s. Developments in the steel industry are thus a harbinger of changes that will eventually pervade the entire economy—especially its traditional sectors. The problems of the steel industry and of the economy as a whole are often related to the upheaval represented by the energy crisis, but this masks the more fundamental structural changes of which the energy crisis is a manifestation rather than a cause. Within the steel industry, the failure to recognize the fundamental processes at work is endemic. The concomitant tendency to seek policies that retard and distort structural changes ensures that the eventual adjustment will be more catastrophic. This is a challenge facing all the major actors in our society—industry, labor, and government.

This book will examine the American steel industry, arguably the most illuminating case study for assessing the implications of the third industrial revolution on traditional, basic sectors of the economy. It will portray a declining industry, but one that is also undergoing a "reindustrial" revolution. The analysis will begin with a treatment of the industry's recent past. It will then examine the steel industry's performance, especially in terms of labor productivity, in the light of international standards, and it will analyze the role of public steel policy since 1950. Finally, it will describe the likely structure of the U.S. steel industry through the rest of this century, and this will define appropriate corporate strategies and governmental policies.

BASIC PROBLEMS IN BASIC INDUSTRIES

Over the last thirty years, the American steel industry has been transformed from a symbol of industrial might into a symptom of industrial blight. Its decline has been protracted and agonizing; neither substantial investment nor the fitful enactment of governmental programs has succeeded in reversing its slide. The industry has suffered one crisis after another over the past twenty-five years—a point borne out by repeated references to the urgency of the industry's condition. The 1970s were "critical years for steel," when the industry had "reached a point of decision."[1] 1977 "marked a major turning point,"[2] while 1980 found "Steel at the Crossroads."[3] Each crisis within the pattern of secular decline has taken its toll; none has been followed by lasting resurgence.

As an industry that was already weak, steel has suffered disproportionately from the 1981-82 recession. Domestic market share has fallen to unprecedented levels while the market itself has shrunk, so that in 1982 the U.S. steel industry shipped less steel than at any time since 1949. Over 50 percent of the industry's capacity stood idle in 1982—a condition not encountered since the 1930s.[4] The present crisis will almost certainly provoke substantial retrenchment and facility closures. There is probably little need to describe what such contraction will mean to the local economies of steel-producing regions or to the steelworkers who will be laid off. The present steel crisis is particularly severe; the recovery from it is certain to be far more sluggish than in the past.

This crisis has attracted considerable governmental attention, partly because of job losses and partly because the festering steel trade problem has aggravated strains with trading partners. Contemporary efforts to address "the steel problem," however, are only the latest in a long series. Through the first half of the postwar period, the interaction between the government and the steel industry was highly antagonistic: the steel problem then centered on the monopoly power inherent in the industry's oligopolistic structure and behavior. By the late 1960s, however, the government's attitude toward the steel problem had shifted from vigilance against the industry's strength to concern about its weakness. This has provoked a plethora of government studies and an ineffectual series of programs concentrating on trade: Voluntary Restraint Agreements, Trigger Price Mechanisms, Tripartite Committees, and "Arrangements Concerning Trade in Certain Steel Products"—the euphemism for the import quota negotiated with the European Community in late 1982.

Recent steel policies have benefited neither the industry nor society as a whole; steel policies have instead been one element in the milieu that perpetuates decline. Fixation on the import question has drawn attention away from the more significant problems facing the industry. No governmental actions have affected the factors which are undermining the competitiveness of American steel producers: anemic market growth, low profits, lagging investment, and high costs. Since the mid-1970s, these problems have afflicted not only the U.S. steel industry but many of its international competitors as well. In Europe, the response has been massive governmental subsidization, and this is the crux of the U.S. industry's complaints about unfair imports. The present depression throughout the world steel industry indicates that the competitive problems facing U.S. producers are not all of their own making. Nevertheless, this is neither a source of solace in the industry's present distress nor a source of hope in its future.

One could be more sanguine about the decline in the U.S. steel industry were it not for the fact that several of the basic industries that have formed the backbone of the U.S. economy are undergoing similar secular declines, characterized by stagnating markets, heightened international competition, inadequate investment, and depressed profits. The decline has been even more precipitous in the U.S. automobile industry, pushing Chrysler to the brink of bankruptcy. Industries that supply the auto industry have also been infected by its disease. Markets have held up fairly well for the machinery industry, but imports have nonetheless made substantial inroads, especially in the computerized segment which increasingly dominates this market. Similar points could be made about most of the traditional manufacturing industries in the United States. The U.S. economy is undergoing a historic shift away from the basic manufacturing industries that have dominated and defined U.S. economic growth since the turn of the century. In the case of steel, this shift began as far back as the 1950s, when growth in steel consumption slowed, and was clearly underway by the 1970s.

Many would argue that such shifts are normal and are no cause for alarm. Market economies, so the argument goes, are inevitably characterized by uneven rates of growth and by an incessant process of decline and renewal—a phenomenon observed not only within industries but among them as well. Under "normal" circumstances, this process operates underneath the surface, with more or less neutral results for the economy as a whole. As the structure of the economy gradually shifts due to changing patterns of consumption and technical progress, the development of new industries compensates for the decline of the old. As basic industries like steel decline, the communications and data processing industries enjoy spectacular growth. The United States is transforming itself into a service economy, we are told, and this is not an inherently negative shift.

Nonetheless, the process of decline and revitalization is not necessarily peaceful or painless. Schumpeter, one of the few economists to pay adequate attention to this pattern, referred to the forces at work as "creative destruction"—a phrase that hardly conveys an image of tranquillity.[5] Unfortunately, the present condition of U.S. basic industries provides ample evidence of destruction, making the creative process of renewal seem paltry by comparison. Growth sectors will have difficulty absorbing the labor force released from declining sectors, and the real income of displaced workers will drop drastically. Since 1973 U.S. wages have declined in real dollars and in 1982 were at their lowest level since 1961.[6] Even if such intimations of decline are actually the price of renewal, the transition will require substantial demographic shifts and the retraining of thousands if not millions of workers.

Faced with the problems of America's basic industries, economists have tended to stress the regenerative effects of reliance on the market and the dangers of governmental interference. Management and labor, with incomparably greater political clout, have insisted on the implementation of measures to protect and strengthen declining industries. Governmental policies have combined elements of both positions. The standard response has been the implementation of stopgap measures which have prevented collapse (e.g., the Chrysler loan guarantees) but which have not represented the sort of coordinated effort that could reverse the decline. This implies that the fundamental, inertial tendency of the U.S. government is noninterventionist. At the same time, however, the political exigencies of industrial decline produce a constant stream of programs designed to defend American industries. Such programs often take the path of least resistance by targeting the trade area while avoiding more drastic measures.

Within the barren landscape of U.S. industrial policy, steel stands out. There is no better test case than steel for the question of how the competitive problems of U.S. basic industries should be interpreted and solved. The secular pattern of decline is clearer in steel, even if it has been less abrupt than in other industries, for example, automobiles. Furthermore, the patterns of response to this decline, both corporate and governmental, have gone through several stages in the steel industry. Throughout the entire postwar period, "what to do about steel" has been a surrogate for the entire question of industrial policy. One purpose of this book is to provide a detailed account of the performance of the U.S. steel industry over the past thirty years and to evaluate governmental programs in that light. While steel provides the content for our analysis, its theme is of wider relevance, namely, the entire issue of sectoral performance and industrial policy.

THE SEARCH FOR A STRATEGY

There is much evidence that governmental policy plays a fundamental role in determining the environment in which industry operates, and this in turn structures industrial performance. Yet too much attention can be given to policy, especially if this distracts attention from the responsibilities of management. The policy environment is largely a datum for management, which must then develop a strategy appropriate to the conditions that determine the prospects of a firm. In the final analysis, industrial performance is determined by the appropriateness of corporate strategy.

The basic elements of a successful strategy are defined by the familiar measures by which performance is judged: profits, sales, costs, investment, and so on. These measures are fairly static, however, and may be misleading in an environment characterized by the kinds of rapid structural and technological changes that now confront the steel industry and other basic industries. Under such circumstances, the demands on corporate strategies are more complex. Competitiveness is defined dynamically, and those companies or industries that prosper are those that have anticipated dynamic changes and positioned themselves accordingly. If we are now undergoing an industrial revolution, companies are confronted with an almost biological imperative: adapt, migrate, or die.

Corporate strategy is a theme that pervades this book. Judged by this standard, the performance of the U.S. steel industry in the postwar period has been poor. In part, this stems from the deeply entrenched oligopolistic traditions of the industry. Faced with new competition—either from other suppliers of steel or from competing materials—the industry has groped in vain for a strategy that could reverse the erosion of its position. It has been mesmerized by nostalgia for its former dominance and has thus failed to accept that, in order to reverse its slide, radical changes are required.

This failing has been common to many U.S. industries. Here again, however, steel is a highly appropriate case study. Its product is relatively homogeneous, and it is less subject to the vagaries of consumers' tastes that affect the performance of many other industries (e.g., automobiles). Since the steel industry has been losing competitiveness since the 1950s, there is a longer record of strategic responses to this phenomenon than is the case elsewhere. Finally, the steel industry has already undergone significant structural changes, so that there is real evidence of renewal within the industry itself. Thus, an embryonic strategy, well suited to the forces affecting the industry, has already emerged.

Successful corporate strategies must be designed to exploit advantages and, more importantly, to turn liabilities into strengths. As we will show, this is an apt description of the strategic orientation of the highly successful Japanese and Canadian steel industries during most of the postwar period; it also applies to the competing "mini-mill" sector that has grown up within the American industry. Firms in the traditional integrated sector, however, have failed to develop a strategy appropriate to their actual conditions. Faced with a seemingly inexorable decline in competitiveness, they have sought to maintain all their facilities. They have linked their future to the availability of massive funds for investment, even as the low profitability of the industry has increasingly constricted their access to capital. Integrated firms in the United States now face the necessity of catching up, and this defines their strategic orientation. Yet this goal has been defined statically. Even in the unlikely event that they are able to quickly reach the performance standards now set by their competitors—domestic and foreign—there is little doubt that their rivals will have made similar progress, thus perpetuating the present competitive disadvantage. The attempt to boost overall performance without undergoing fundamental structural changes is therefore doomed. A more innovative and yet more realistic strategy is required. To a great extent, the basic purpose of this book is to suggest the contours of a more appropriate strategy for the industry and for integrated firms. This theme dominates the final chapter.

A SYNOPSIS

Effective strategies and policies must be based on long-term rather than immediate problems and possibilities. From this perspective, what is happening within an industry is more important than what is happening to it; sophisticated historical understanding is needed to tell the difference. Part I of this book broadly describes the postwar history of the American steel industry. This history has a distinctly schizophrenic character. The industry's present problems are so pervasive that it is difficult to grasp the extent to which it was the world leader in the first decade after World War II. As is described in Chapter 2, the U.S. position in the world steel industry in the early 1950s was akin to Gulliver's status among the Lilliputians. Viewed retrospectively, some of the causes of the industry's subsequent decline can be located in the immediate postwar period, but this does not mean that they were apparent at the time. Since then, the U.S. steel industry has been forced to retreat on almost every front; its technological lead has effectively disappeared, profits have consistently failed to reach the manufacturing average, and market share has eroded in the face of imports from increasingly diverse sources. After 1960, the evidence of decline was unavoidable, and this is presented and analyzed in Chapter 3.

The decline has not been an orderly retreat–a fact masked by the dryness and regularity of its statistical depiction. Instead, it has been a process of structural change, the tempo of which has increased over time. Some of the forms taken by this structural change have been concentrated in the integrated sector of the industry, and these are by now fairly familiar: bankruptcies, closure of facilities, reductions in capacity, and increasing diversification away from steel production. A less recognized structural change has been the spectacular growth of the so-called mini-mills. This sector–which is generally characterized by much smaller plants, reliance on local markets, and scrap-based electric furnaces–has grown from an insignificant to a major role in the U.S. steel market. In terms of profitability and international competitiveness, its performance has been the reverse of its integrated counterpart. Several minimill firms are characterized by Wall Street analysts as high-tech, high-growth firms–surely a surprising depiction for those who are accustomed to portrayals of the steel industry as a unitary whole.[7]

It is the relationship between these sectors that best defines the prospects for the revitalization of the American steel industry. Without a thorough understanding of this relationship, there is little chance that policymakers or, for that matter, steel executives will be able to reverse the industry's competitive slide. While Chapters 2 and 3 focus on the traditional integrated sector, Chapter 4 discusses the great contrasts in performance that are masked by the statistical aggregates referring to *the* steel industry as though it were homogeneous.

Part II deepens this analysis by presenting a specific account of comparative postwar trends in labor productivity–the most revealing of the various factors that define or determine industrial performance. Chapter 5 presents a detailed comparison of labor productivity trends–by process and by product—in both the integrated and mini-mill sectors of the Japanese and American steel industries. The patterns described there are interpreted in Chapters 6 and 7. Chapter 6 focuses on the sources (market growth, investment, technology, etc.) of the record described in Chapter 5. Chapter 7 uses this analysis to criticize the strategy through which American producers now pursue improved performance and to suggest an alternative. This alternative is defined in terms of capital costs as well as labor productivity and concentrates on the relative importance of scale and technology.

Chapters 8 and 9 discuss the policy environment; together with economic conditions and managerial strategies, this determines the performance of an industry. The U.S. has had little experience with explicit industrial policies; for that reason, we first turn to the lessons provided by the steel policies of Japan and Canada. Canada is a particularly interesting case, since the cultural and political traditions of that country are so close to those of the United States. Chapter 9 describes the sorry history of relations between the U.S. government and the American steel industry. Paradoxically, this relationship has been characterized not only by a widely recognized mutual hostility but also by a less apparent convergence of interests, based on resistance to the competitive forces which are reshaping the industry.

Finally, Chapter 10 will reconsider the arguments raised in the preceding sections of this volume in order to delineate the optimal structure of the U.S. steel industry for the rest of this century. The speed and ease with which the inevitable transition occurs will be determined by corporate strategies and governmental policies. Chapter 10 also describes the strategy that should be pursued by integrated firms in the United States and the themes that should control the formation of steel policy. While contraction is certain, the right mix of strategy and policy—one which embraces the inevitable and thus turns it to advantage—can ensure that the survivors no longer show symptoms of the disease.

Notes

1. William T. Hogan, S.J., *The 1970s: Critical Years for Steel* (Lexington, Mass.: D.C. Heath, 1972), p. 1.
2. Robert Crandall, *The U.S. Steel Industry in Recurrent Crisis* (Washington, D.C.: The Brookings Institution, 1981), p. 1.
3. American Iron and Steel Institute (AISI), *Steel at the Crossroads* (Washington, D.C.: AISI, 1980).

4. AISI, *Annual Statistical Report* (New York and Washington, D.C.: AISI, various years).
5. See, for example, Joseph A. Schumpeter, *Capitalism, Socialism, and Democracy* (New York: Harper Torchbooks, 1962), pp. 81-87.
6. Calculated from data provided in *Economic Report of the President, 1983* (Washington, D.C.: U.S. Government Printing Office, 1983), p. 207.
7. See, for example, Kenneth P. Pearlman, "Emerging Growth Companies," *Emerging Growth Stocks*, 2, no. 3 (New York: Shearson American Express, August, 1981).

2 POSTWAR PREEMINENCE, 1950–1960

Americans, of course, don't like to take second place in any league, so they expect their steel industry to be bigger and more productive than the steel industry of any other nation on earth. It is; but what many Americans do not know is that their own steel industry is bigger than those of all the other nations on earth put together.
No other nation in the world could have matched that record. It is a record that stands as a glorious tribute to the men who make steel and the men who built steel in America.[1]

—Benjamin Fairless
Chairman, U.S. Steel Corporation
January 1951

In 1950, over 45 percent of the world's raw steel was produced in the United States. Admittedly, this awesome status was an anomaly, for it was based in part on the wartime destruction of steelmaking facilities outside the United States. Yet reconstruction, in Europe at least, was largely complete by this date, so that the stature of the U.S. industry was not solely due to the decimation of its foreign counterparts. The U.S. advantage was solidly grounded in the technological parameters of steel production. American steel mills at that time were generally the biggest, the newest, and the most technologically advanced in the world. In the early 1950s, the U.S. industry's eventual competitors showed little evidence of their future prospects, as is indicated by the extent to which U.S. output dwarfed that of other countries (Figure 2–1).

Decades later, it is difficult to recall the dominance enjoyed by the U.S. steel industry during the immediate postwar period. We have become accustomed to portrayals of today's steel industry as plagued by severe competitive problems, faced with persistent decline, and dependent on governmental favors. Yet the competitive strengths and advantages of the U.S. steel industry were very evident as late as the 1960s

Figure 2–1. Growth in Raw Steel Production: Selected Countries, 1950–1981 (3-year moving averages).

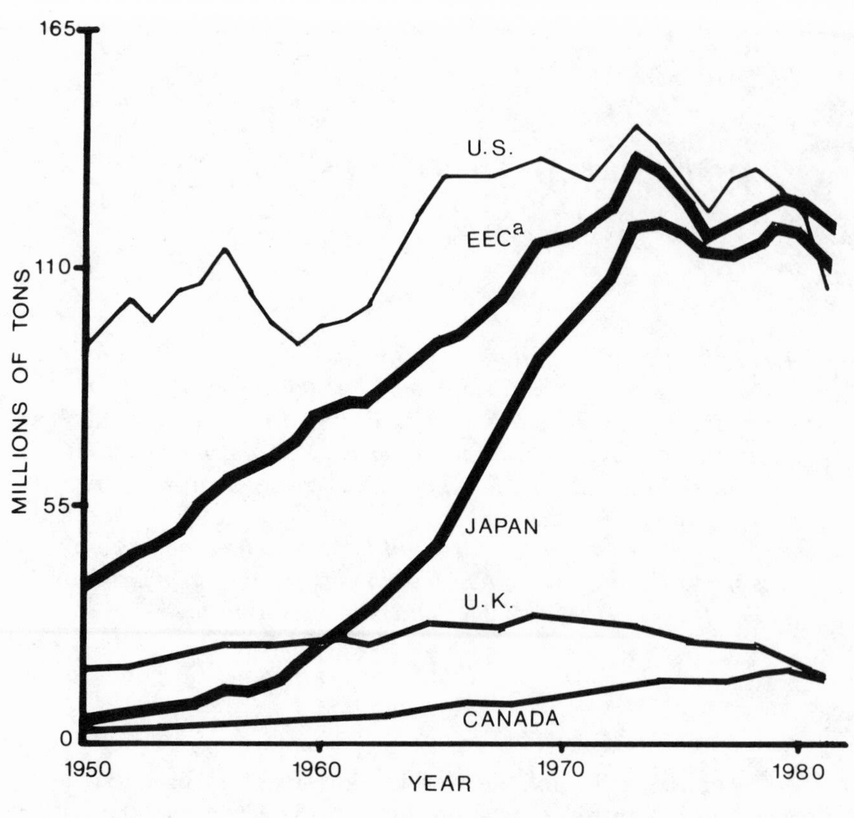

a. Original Six (Benelux, Germany, France, Italy).
Sources: Federal Trade Commission, *The U.S. Steel Industry and Its International Rivals* (Washington, D.C.: U.S. Government Printing Office, 1977); International Iron and Steel Institute (IISI), *Statistical Yearbook* (Brussels: IISI, various years).

and still have not completely disappeared. In the early 1950s, its dominance was touted by both the industry and its critics. In fact, the principal challenges faced by the industry at that time came not from foreign competitors but from critics in government and academia.

The international status enjoyed by the American steel industry during the early postwar period was common to many basic industries in the United States. Several of the key parameters of international competitiveness did not favor American industries at that time. In steel, for example, the extent to which U.S. wage rates exceeded those of its international competitors was relatively greater in the first fifteen years after World War II than at any time before or since. Nonetheless, such disadvantages were overwhelmed by the technological superiority and outstanding productivity of U.S. firms. As subsequent events have shown, U.S. industries were unable to maintain this technological advantage; this is a key theme in the declining competitiveness of American steel. This chapter, however, concerns a period when the signs of decay were faint, when the competitive prowess of American steel seemed unchallenged.

TECHNOLOGICAL SUPREMACY AND ITS EFFECTS

While the output levels described in Figure 2–1 indicate the global dominance of the American steel industry in the early postwar period, more detailed comparison suggests that these data actually understate the U.S. lead. Japanese and European competitive disadvantages were evidenced not only by low output levels but, more importantly, by the relative backwardness of steel facilities in these nations. Thus, the quantitative predominance of the United States was bolstered by qualitative advantages that, although difficult to measure, were equally impressive. Through the mid-1950s, the available indicators of production efficiencies and technological performance uniformly favored the United States.

The standard procedure for the production of steel in the 1950s involved several discrete processes, and with some variations these processes are the same today (see Appendix A). Raw materials are prepared: iron ore can be agglomerated to provide a more concentrated charge, and coal is baked to produce coke. These raw materials are combined in a blast furnace, where the iron ore is reduced to produce molten iron. (The above processes are sidestepped when the electric furnace, which generally relies on scrap rather than on iron, is used.) The carbon content of the molten iron is then reduced in the steelmaking furnace. Finally, the semifinished shapes that are the end product of the steelmaking process are heated and rolled into finished shapes (bars, sheets, structural shapes, rods, plates, etc.). One method of weighing the relative efficiencies of various steel industries is to compare the extent to which the best techniques are applied for each of these processes; another is to compare the scale of operation for these processes and for firms and plants as a whole. Both of these tests will be applied below.

Table 2–1. Use of Advanced Technologies in the Mid-1950s.

	U.S.	*U.K.*	*W. Germany*	*France*	*Japan*
A. Preparation of materials, 1958					
Share of sinter in the blast furnace burden (%)	40	35	24	8	53
Production of ferrous agglomerates (mill. tons)	35.3	9.6	11.0	2.7	6.4
B. Raw Steel Production, 1954 (mill. tons)	88.3	20.7	19.2	11.7	8.5
C. Share of output by furnace type, 1954 (%)					
Open hearth	91.0	87.7	56.2	32.0	82.1
Electric	6.2	5.0	3.9	7.8	13.2
Other	2.9	7.3	39.9	60.2	4.7

Sources: U.N., *Long-term Trends and Problems of the European Steel Industry* (Geneva: United Nations, 1959); Japan Iron and Steel Federation (JISF), "Statistical Yearbook for 1960" (Tokoyo: JISF, 1961); and Anthony Cockerill, *The Steel Industry; International Comparisons of Industrial Structure and Performance* (Cambridge; Cambridge University Press, 1974).

Superior Processes

Table 2–1 presents various data on the historical application of superior processes, concentrating on two stages: the preparation of iron ore—especially via pelletizing and sintering—and steelmaking. For several reasons, improving the quality of the iron ore burden (i.e., the ferrous input reduced in the blast furnace) was a prime technological goal in the immediate postwar period. The best domestic ore fields were being exhausted in both the United States and Europe, so that inferior ores were becoming more prevalent. This placed severe burdens on ironmaking efficiency. The sintering procedure, a steel mill operation that agglomerates ore fines into a more usable form, was one response to this trend. Others were increased efforts to beneficiate ore at the mine sites through pelletization and an intensified search for high-grade deposits in nonindustrial areas. Table 2–1 shows that, with the exception of Japan (whose overall capacity was small at the time), the U.S. share of sintered material in the blast furnace burden exceeded that of all major competitors in 1958. U.S. production of agglomerated iron materials (pellets, sinter, etc.) far outstripped the levels achieved by its major competitors. During the 1950s the United States maintained the technological lead in the ore beneficiation processes that were being widely adopted at that time.

For the steelmaking step, the preferred technology during most of the 1950s was the open hearth. Open hearth technology has since been replaced as the most efficient integrated process by the basic oxygen technique. Early basic oxygen furnaces were quite small, however, so that their commercial adoption was not widespread until the late 1950s. Table 2–1 shows the relative shares of different steelmaking techniques in total steel production for various countries as of 1954. At that time, electric furnace technology was predominantly used for the production of specialty steels, so that it competed with open hearth techniques only minimally. The significant comparison for steelmaking efficiency is between the open hearth and converters (either Bessemer or Thomas). Table 2–1 shows that in 1954 the United States had the highest open hearth share among the countries compared, while its use of inefficient processes was the lowest. In Europe, over one-third of total steel output was produced in relatively inefficient Bessemer or Thomas converters. European reliance on such facilities was especially prevalent in mills that used French ores, which are too rich in phosphorus to be refined efficiently in open hearths. Nonetheless, the fact that the continued use of what was basically a nineteenth century technology can be explained does not reduce the competitive disadvantage inherent in the choice of the outmoded technique.

Thus, by the mid-1950s the technological performance of the American steel industry was clearly superior to that of its major international rivals in two key process steps: raw materials preparation and steelmaking per se. Unfortunately, the choice-of-technique criterion for evaluating technological performance is not easily applied for this period to other processes in the production of steel. Technological differences in ironmaking, rolling, and so on are more easily discerned in the scale of operations than in different techniques.

Scale

Because of its capital intensity and its chemical and physical complexities, steel production is characterized by significant economies of scale. Table 2–2 presents evidence on the scale of steel plants and firms for the

Table 2–2. Average Capacity of Large[a] Plants and Firms in 1954.

	U.S.	*U.K.*	*Japan*	*France*	*Canada*
A. Plants					
Average capacity[b]	2.76	1.10	.75	1.12	1.44
Percentage of U.S.	100	40	27	41	52
B. Firms					
Average capacity[b]	10.34	1.84	1.50	1.96	1.44
Percentage of U.S.	100	18	15	19	14

a. Counting only the largest plants and firms up to two-thirds of total national capacity.
b. Millions of net tons.
Source: A. Cockerill, *The Steel Industry: International Comparisons of Industrial Structure and Performance* (Cambridge; Cambridge University Press, 1974).

United States and several other countries in the mid-1950s. The scale advantage of the U.S. industry is starkest when judged in terms of firm capacity. In none of the other countries was the average firm even 20 percent as large as the average firm in the United States. Doubts can be raised about the significance of firm-based economies of scale, however, since they lack the technological foundation of plant-based economies. This is not the case for plant-based economies of scale, which have significant economic effects. While the U.S. advantage in plant size is not as stark as in firm size, it is nonetheless impressive. Even leaving out many small firms by eliminating the smallest firms up to one-third of total capacity, in the mid-1950s only Canada, a minor player in the world steel market, reached an average plant size half that found in the United States—and then only barely.

The small scale of foreign steel industries in the mid-1950s represented a significant economic disadvantage. In 1956, J.S. Bain estimated that the minimum efficient scale for an integrated steelworks fell within a range of 1 to 2.5 million tons of annual crude steel capacity.[2] Even accepting the lower pole of this range as denoting minimum efficient scale, over 50 percent of total steelmaking capacity in France and Britain failed to attain the minimum size required for competitiveness. Over 75 percent of Japanese plants were smaller than this one-million-ton standard, while almost 80 percent of all U.S. plants exceeded Bain's minimum efficient scale.[3]

As one would expect, the preeminence of the U.S. steel industry in terms of plant scale also applied to the scale of individual facilities. This is shown in the comparison presented in Table 2–3, which refers to three countries only—Japan, the United States, and Canada. This table covers

Table 2–3. Average Capacity of Selected Facilities in 1958: U.S., Japan, Canada (thousands of net tons per annum).

		U.S.	*Japan*	*Canada*
A.	Blast furnaces	342	300	262
B.	Open hearths	127	77	136
C.	Hot strip mills	1371	660	930
D.	Plate mills	363	127	428[a]
	Average facility size[b]	387	220	249

a. Canada had only one plate mill in 1957.
b. Weighted by facility share of total calculated capacity.
Sources: American Iron and Steel Institute (AISI), *Directory of Iron and Steel Works in the U.S. and Canada* (New York: AISI, 1957); Japan Iron and Steel Federation (JISF), "Statistical Yearbook for 1960" (Tokyo: JISF, 1961).

the entire range of steelmaking operations, from the production of iron (blast furnaces), through steelmaking itself (open hearths), to rolling and finishing (hot strip mills and plate mills). Weighting each of these facilities by total capacity, the U.S. advantage is quite significant. Comparing the weighted averages of facility scale as presented in Table 2–3, these data suggest that the scale of U.S. facilities was on average 76 percent greater than in Japan and 56 percent greater than in Canada. In certain cases, the Canadian industry had a scale advantage over the United States. This was due, however, to the small size of the Canadian steel industry (for instance, the Canadian industry possessed only one plate mill at that time).

The average scale of European facilities, at least in the hot end of the steelmaking process, was also well below U.S. norms. In 1953, for instance, average annual blast furnace capacity was 308,000 tons in the United States, about 130,000 tons in Great Britain, and less than 175,000 tons in West Germany. For the open hearth furnace, average heat size in 1953 was 140 tons in the United States, less than 100 tons in Great Britain, and less than 50 tons in France.[4]

Cost Effects

Advanced technology and significant scale economies enabled the American steel industry to set the world standard for physical efficiency and productivity, an advantage that in turn provided the industry with relatively low operating costs. The impact of these physical advantages on comparative costs is suggested by Table 2–4, which compares U.S. and Japanese operating costs for the production of cold-rolled sheet, a highly representative integrated product. While this comparison excludes interest, taxes, and other financial costs, it nevertheless reflects the overall cost competitiveness of the two industries.

Two major areas of difference are illustrated in Table 2–4—input prices and labor productivity. In regard to the former, the data show that the Japanese faced a substantial disadvantage in terms of the cost of raw materials. The prices paid by the Japanese for basic raw materials were on average 1.5 times those found in the United States—a disadvantage that stemmed largely from Japan's dependence on imported raw materials. Surprisingly, in spite of the U.S. industry's technological advantage, the Japanese industry appears to have almost matched its U.S. rival in terms of the efficiency with which material inputs were used. As far as raw materials are concerned, therefore, the Japanese disadvantage was based primarily on price rather than on efficiency. Recognition of this fact was a crucial component in the Japanese industry's eventual success. New plants were constructed at tidewater sites, massive ore and coal carriers were built to reduce unit shipping costs, and long-term contracts for high-quality ore and coal were signed with several foreign sources. All of these efforts had the result of reducing the delivered price of materials and thus the Japanese disadvantage on that front.

Table 2–4. Estimated Operating Costs for Production of Cold-rolled Sheet in 1958, U.S. versus Japan.

	Price of Input		*Use/Net ton CRS*		*Unit Cost ($)*	
	U.S.	*Japan*	*U. S.*	*Japan*	*U. S.*	*Japan*
Labor	$3.75/hr.	$0.58/hr.	11.58	36.65	43	21
Iron ore	$10.64/t.	$14.73/t.	1.44	1.47	15	22
Purchased scrap	$34.07/t.	$43.25/t.	0.20	0.20	7	9
Coking coal	$10.50/t.	$19.35/t.	1.04	0.98	11	19
Other energy					14	20
Other[a]					24	37
		Total materials (+ other)			71	107
		Total operating costs			114	128

a. Refractories, rolls, fluxes, alloying agents, etc.

Sources: Federal Trade Commission, *The U.S. Steel Industry and Its International Rivals*, (Washington, D.C.: U.S. Government Printing Office, 1977) and authors' estimates.

The main difference between the Japanese and American steel industries, however, was in labor productivity. U.S. labor requirements were less than one-third the Japanese, providing the American industry with an enormous advantage. Unit labor costs were still twice as great in the United States as in Japan, but only because wages and benefits were over six times as high—a graphic indication of the welfare effects of high productivity. Given the enormous U.S. lead in productivity, Japanese steelmakers maintained some measure of price competitiveness only on the basis of extremely low employment costs. A similar argument would apply to the competitive prospects of all major steel industries vis-a-vis the United States.

This combination of outstanding physical efficiency and high relative wages was characteristic of the U.S. manufacturing sector in the immediate postwar period. Steel is a particularly good example of this, both because of its importance within any industrialized economy and because its relatively homogeneous character makes it possible to construct meaningful international comparisons. Table 2–4 illustrates the cost advantage that defined America's postwar preeminence in steelmaking. The even more substantial U.S. advantage in technology and scale was a cushion, offsetting disadvantages (actual or potential) in other areas, particularly relatively high employment costs. In analyzing the different status of the industry in the 1970s and 1980s compared with the 1950s, the erosion of this cushion and the broader implications of its erosion must be given a prominent place. U.S. producers enjoyed a static cost advantage in the 1950s, but this did not necessarily imply a corresponding dynamic advantage in terms of cost trends or technological progressiveness—notwithstanding the common belief that international leadership was the manifest destiny of American industry. Sales of U.S. technology encouraged foreign producers to adopt American techniques, and this gradually closed the efficiency gap which had formerly set the U.S. industry apart. Foreign producers' adoption of American technology should have been warning enough that the U.S. advantage was vulnerable. In the first fifteen years after World War II, however, the industry's technical supremacy seemed secure. Moreover, it had proved in preceding decades that its ability to invest rapidly in the most up-to-date technology was unparalleled. At that time, there was little indication that this status could be lost.

DIVERGING TRENDS IN CAPACITY AND CONSUMPTION

The technological supremacy of the U.S. steel industry in the immediate postwar period, for all its seeming invulnerability, was not a result of natural law. Rather, it was an economic fact that must itself be explained; the principal explanation lies on the demand side of the economic equation. As with many other American industries, the steel industry had long benefited from the growth and extent of the U.S. market. The large domestic market had permitted American firms to size their operations efficiently, an advantage that was unattainable by other producers so long as their markets were largely limited by national boundaries. Until the 1930s the U.S. steel market had grown rapidly enough to ensure that additional capacity, embodying the latest technology, would be built to match market growth. While this was interrupted by the depression, defense needs during World War II had required significant new capacity, most of it built at government expense. Despite the aftereffects of the depression, World War II left the U.S. industry with expanded and modernized capacity.

Contention over Capacity

Nonetheless, the industry entered the 1950s with a great deal of uncertainty about future demand trends and their effects on the investment plans and financial performance of individual firms. Caution was an understandable attitude for steel executives who had lived through the disastrously low operating rates of the 1930s. The weight given to avoiding excess capacity was expressed in the quotation that opened this chapter. While that statement projected a tone of confidence and vigor, its underlying theme was the historical persistence of excess capacity and the adequacy of the industry's productive potential as it entered the 1950s.

In fact, U.S. Steel's chairman made those comments in the context of an intense controversy over the adequacy of U.S. steel capacity. In the late 1940s and early 1950s, the U.S. government pressured the industry to raise its capacity (see Chapter 9). History demonstrates the correctness of the industry's contention that adequate capacity existed through the 1950s. While production exceeded rated capacity in 1951, the year in which this dispute was most intense, it did not exceed 95 percent of capacity in any succeeding year.[5]

Far from having expanded to an inadequate extent in the 1950s, the U.S. steel industry actually increased capacity substantially in what was essentially a flat market (see Table 2–5). The overall trend in demand growth for the 1950s did not in fact represent a definitive break with the stagnation characteristic of the immediate postwar period. This is shown in Table 2–6 and Figure 2–2, where trends in U.S. steel consumption (apparent consumption or apparent supply equals net shipments plus imports minus exports) are compared with those in several other countries or regions. These data show the anemic growth of U.S. steel

Table 2–5. Growth in U.S. Steel Capacity, 1950–1981 (compound annual percentage rates).

1950–55:	4.48	1965–70:	0.36
1955–60:	3.23	1970–75:	0.13
1960–65:	0.49	1975–81:	-0.24

Sources: Council on Wage and Price Stability (COWPS), *Prices and Costs in the U.S. Steel Industry* (Washington, D.C.: U.S. Government Printing Office, 1977); American Iron and Steel Institute (AISI), *Annual Statistical Report, 1981* (Washington, D.C.: AISI, 1981).

consumption through the 1950s. Since capacity grew at a compound annual rate of over 4 percent while consumption grew at 0.4 percent, one can surmise that the expansion of the U.S. steel industry during the 1950s was based on an overestimation of potential demand growth. This unjustified optimism had the expected effect on operating rates, which averaged 89 percent for 1951–55 and 73 percent for 1955–60. There is little doubt that the demands of the Korean War boosted capacity utilization in the first half of the decade or that the 1958 recession and the 1959 strike reduced it in the latter half. Yet these phenomena only camouflaged the underlying trends in capacity and domestic demand.

International Comparisons

Figure 2–2 and Table 2–6 also show that during the 1950s market growth in other regions, especially the European Economic Community (EEC)

Table 2–6. Growth in Apparent Steel Consumption (compound annual percentage rates, 1950–1981).[a]

Period	*U.S.*	*Japan*	*Canada*	*U.K.*	*EEC*[b]
1950–60	0.4	17.3	2.5	3.3	8.3
1960–69	4.3	13.1	6.8	2.5	5.6
1969–81	−0.9	1.3	1.6	−3.5	−0.9
1950–81	1.0	9.8	3.1	0.3	3.6

a. Calculated on a crude steel equivalent basis from three-year averages.
b. Original Six (Benelux, Germany, France, Italy).
Sources: Federal Trade Commission (FTC), *Staff Report on the U.S. Steel Industry and Its International Rivals* (Washington, D.C.: U.S. Government Printing Office, 1977); International Iron and Steel Instute (IISI), *Statistical Yearbook* (Brussels: IISI, various years).

Figure 2–2. Growth in Apparent Consumption: Selected Countries, 1950–1981 (crude steel equivalent, 3-year moving averages).

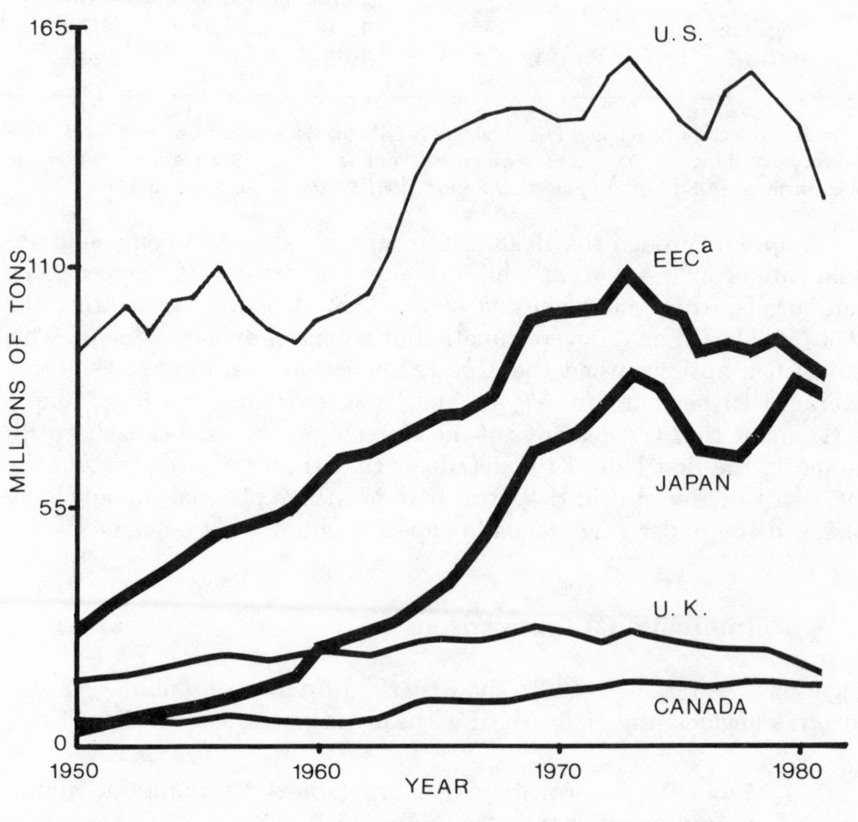

a. Original Six (Benelux, Germany, France, Italy).
Sources: Federal Trade Commission (FTC), *The U.S. Steel Industry and Its International Rivals* (Washington, D.C.: U.S. Government Printing Office, 1977); International Iron and Steel Institute (IISI), *Statistical Yearbook* (Brussels: IISI, various years).

and Japan, far outstripped that in the United States. In the case of Europe, at least, this represented a reversal of the pattern of previous decades. Steel consumption in Western Europe had grown at an annual rate of 1.5 percent from 1913 to 1929 and 0.9 percent from 1929 to 1951. The comparable figures for North America were 3.1 percent and 2.6 percent respectively.[6] The Japanese market had grown as rapidly during these periods, but it was only one-tenth as large as its American counterpart. Hence, the Japanese market was still inadequate to encourage the realiza-

Table 2–7. Growth in Raw Steel Production (compound annual percentage rates, 1950–1981).[a]

Period	*U.S.*	*Japan*	*Canada*	*U.K.*	*EEC*[b]
1950–60	0.4	16.6	5.9	3.5	7.9
1960–69	3.8	15.4	7.1	2.1	4.6
1969–81	−2.1	1.9	2.7	−5.5	0.4
1950–81	0.4	10.3	5.0	−0.4	4.0

a. Calculated from three-year averages.
b. Original Six (Benelux, Germany, France, Italy).
Sources: International Iron and Steel Institute (IISI), *Statistical Yearbook* (Brussels: IISI, various years); U.N. Statistics.

tion of substantial economies of scale. Relatively rapid market growth in what was already the largest market in the world had supported a seemingly inevitable U.S. advantage in scale and technology.

During the 1950s, consumption grew more slowly in the United States than in the other regions included in Table 2–6. The U.S. market rebounded in the first half of the 1960s, but the benefits of this were largely offset by shifts in the pattern of trade. Whereas the U.S. industry had been a net exporter from the late nineteenth century until 1959, U.S. imports exceeded exports thereafter. Thus, much of the demand growth was taken by foreign companies—a point that will be discussed at greater length in the next chapter. For the entire postwar period, the U.S. market grew more slowly than that of any other major steel-producing country except Great Britain.

Anemic market growth had predictable consequences for the trend in the industry's output (see Figure 2–1 and Table 2–7). When compared with Table 2–6, Table 2–7 shows that, for each period, U.S. production grew even more slowly than steel consumption—a result generated by the waning international status of American producers, which caused exports to fall and imports to increase. In Japan and the EEC, however, growth in production more or less matched the growth in domestic consumption through the 1960s, although the link between these variables was broken in the 1970s. Japan provides the most striking example of the dynamic effects of demand growth on international competitiveness; its home market grew at an annual rate of 17.3 percent during the 1950s.

Over the entire period from 1950 to 1981, the Japanese market grew at an annual rate of 9.8 percent, doubling every eight years. Planning investment to cope with market growth of this magnitude means that modern plants of optimal scale can be designed and built—in fact, must be built if supply is to be adequate to demand. For the United States, on the other hand, market growth at the 1950–81 rate (1.0%) would mean that the market would double only after seventy years. In the face of such sluggish market growth, optimally sized new plants will have a significant marginal effect on operating rates, and thus on financial performance, unless some capacity is also retired. This fact vitiates the incentive to modernize and invest, thus exacerbating the competitive disadvantage vis-a-vis an industry that is expanding rapidly.

The slow growth in U.S. steel consumption during the 1950s was an early indicator, if not cause, of the problems that would develop in the 1960s and dominate the 1970s. The market base upon which the U.S. technological advantage was built had ceased to expand—both absolutely and, more importantly, relative to the very vigorous and sustained growth in European and Japanese consumption. The booming demand enjoyed by European and Japanese producers provided them with a constellation of factors similar to that which had previously maintained the dominance of the U.S. industry. In the case of Western Europe, the integration of distinct national economies into the European Community created a market that at least approached that of the United States in terms of size—with salutary effects on the scale of new operations. In the case of Japan, the benefits of domestic market growth were compounded by the Japanese commitment to export steel. Significant reductions in shipping costs, accelerated by the construction of large bulk carriers and tidewater plants, made it possible for Japanese (and other) producers to plan investment for the growth of the international steel market—a fact that is indicated by the extent to which the growth in Japanese production outstripped even the vigorous growth in Japanese consumption.

In the United States, slow market growth did not keep the domestic industry from expanding, possibly in response to governmental concern that the economy be assured of an adequate supply of steel. This expansion had a price, however: the spectres of excess capacity, low operating rates, and reduced profits. Forced to cope with a low-growth environment, the first response of the American steel industry was to invest as though market growth were robust. While this ensured that new facilities were put in place, it also generated pressures that began to threaten the industry's performance in the late 1950s.

INVESTMENT

The effects of weak market growth were compounded by the inappropriateness of the industry's investment effort. Expanded capacity brought problems of its own, but an equally negative aspect of steel investment during the 1950s was the limited extent to which it improved efficiency and performance. Since technological superiority was the key element in the competitive arsenal of American steelmakers, investments that failed to incorporate technological advances weakened the industry's prospects—especially when combined with the disturbing trends in capacity and relative rates of market growth.

Total productive investment in the steel industry averaged 2.6 billion dollars (in 1981 dollars) during the 1950s—a level little different from the depressed levels of the late 1970s.[7] Relative prosperity in the immediate postwar period was therefore not associated with a corresponding commitment to the reinvestment of earnings. The low level of investment during the 1950s is even more surprising given the prodigous and unprecedented expansion achieved in the course of that decade, when capacity rose from 100 million tons in 1950 to 148 million tons in 1959. Despite this, total steel industry investment actually increased significantly during the following decade, even when judged in terms of investment per ton of capacity (see Figure 3–5). Yet capacity grew little after 1960—a fact that suggests that the nature of investment had changed along with its level. Whereas capital expenditures in the 1950s went principally to major expansion projects, subsequent investments have been targeted chiefly at the modernization of existing facilities, at least for the integrated sector of the industry.

Investment in the steel industry during the 1960s and 1970s will be discussed in more detail in the next chapter. During the 1950s, investment was generally of the capital-widening rather than capital-deepening type—that is, funds were devoted primarily to additional productive units embodying familiar technology rather than to the adoption of new technologies. Thus, the investments that fueled the expansion of steelmaking capacity failed to incorporate major technological breakthroughs. Besides the commitment to capital-widening expansion, substantial funds were devoted to the discovery, development, and upgrading of iron ore properties. While this effort involved the adoption of some new technologies, their cost effects were not necessarily advantageous.

The capital-widening character of steel investment during the immediate postwar period is reflected in data on the scale of operations for various facilities. Total blast furnace capacity, for instance, increased from 71.5 million tons in 1950 to 95.7 million tons in 1960—a compound annual growth rate of 3.3 percent. Average blast furnace capacity, however, rose at an annual rate of less than 3 percent.[8] Such retarded growth in the scale of operations is even more evident for coke ovens. In 1951, the annual capacity of the average by-product coke oven was 4,938 tons. By 1960, this had increased to only 5,170 tons, even though total coke capacity had grown from 62.9 to 71.4 million tons. Thus, while coke capacity had increased at a compound annual rate of 1.4 percent from 1951 to 1960, the average size of coke plants had increased at a rate of only 0.5 percent.[9] This indicates that for coke production and ironmaking, steel industry investment during the 1950s had a significantly greater impact on capacity than on scale. This is what one would expect of capital-widening investment.

There is a somewhat different result, however, in regard to the open hearth furnace. This was the principal target of steel industry investment during the 1950s, and here the expansion of capacity is associated with an even sharper rate of increase in the scale of operations. From 1951 to 1960, the open hearth capacity of major U.S. steel producers increased at the robust compound annual rate of 3.7 percent, from 90.4 million tons in 1951 to 125.9 million tons in 1960. In spite of this massive increase in capacity, the number of open hearth furnaces actually dropped from 910 to 874.[10] Thus the annual capacity of the average open hearth furnace rose from 99,302 to 144,013 tons—a yearly increase of 4.2 percent.

During the 1950s, then, steel industry investments increased the scale of open hearths far more than was the case for blast furnaces or coke ovens. As a result, the greatest increases in efficiency were presumably achieved at the steelmaking stage, in the open hearths. This was the opposite of what subsequent technological developments required. The technologies of ironmaking and coke production, where investment was not matched by scale increases, did not change significantly after the 1950s. Advances made in these areas during the 1950s would continue to have their full effects in later years. In the case of the open hearth, however, the improvements attained in the 1950s rapidly became irrelevant, since this technology essentially ceased to be competitive within a few years after the new open hearths of the 1950s were in place. By the end of that decade, the replacement of the open hearth by the basic oxygen furnace had become essential. A few years thereafter, *Business Week* would characterize the capital-widening investment program of the 1950s as "40 million tons of the wrong kind of capacity—the open hearth furnace."[11]

The steel industry's second major investment program of this decade involved iron ore; campaigns were undertaken on several fronts to upgrade the industry's raw materials base. First, efforts were made to increase the quality of the blast furnace burden (i.e., the ore with which the furnace is charged) and to reduce waste. At the steel mills, this involved the construction of sintering facilities. As we have already shown, the United States was the leader in the application of sintering technology through the mid-1950s (see Table 2–1). However, sintering did not address the fundamental deterioration of traditional U.S. iron ore sources in the upper Great Lakes. To deal with this problem, the industry followed a dual strategy, seeking higher quality reserves in other parts of the world and investing in a technology capable of exploiting the massive low-grade deposits (chiefly taconite) in the upper Great Lakes.[12]

In the late 1940s and 1950s, U.S. firms discovered substantial iron ore reserves outside the United States. Although investments were made in South America and Africa, Canada—more specifically, Quebec and Labrador—was the preferred site for most U.S. firms. In hindsight, this Canadian orientation had disappointing results; deposits in other countries—e.g., Brazil and Australia—would prove to be larger, richer, and better located for export (see Appendix B). Several factors made Canadian sourcing more attractive: Canadian reserves were closer to most U.S. mills, and the political risks were lower in Canada than outside North America. Nonetheless, companies in Japan and Europe increasingly gained a cost advantage through their reliance on iron ore reserves of better quality than those available in North America. This advantage was strengthened by the relatively difficult logistics of Canadian supply and the increased use of bulk carriers in the international ore trade. Oceanborne shipments of iron increased their share of total world iron consumption from 16 percent in 1950 to 23 percent in 1960 and 38 percent in the mid-1970s, a level they have held since then.[13] The United States has been only marginally involved in this development; iron ore (or pellets) from outside North America accounted for less than 15 percent of U.S. consumption in the late 1970s—a lower share than in the late 1950s, when it exceeded 20 percent.[14]

U.S. firms also made efforts to develop a technological means of coping with the deteriorating quality of domestic sources of iron, particularly in the upper Great Lakes. Low-grade taconite deposits (20 to 35% ferrous content) were abundant, although they required beneficiation before they could be consumed in steelmaking operations. Beneficiation was carried out at mine site pelletizing plants, using a technology which began to be applied in the late 1950s. Pelletizing is now used extensively even at sites where the iron ore is of high quality, since it simplifies transport and provides a more suitable blast furnace burden.

Pelletizing plants that are designed to upgrade low-quality taconite ores involve tremendous capital expenditures. The scale of investment required for major pelletizing plants is indicated by the cost of the largest U.S. pelletizing plant built in the 1950s. Located near Duluth, it became fully operational in 1958. With an annual pellet capacity of 10.3

million tons, it was built by a consortium of U.S. and Canadian firms at a cost of $300 million—close to one billion 1981 dollars. This figure is more than half the cost of the entire Benjamin Fairless Works of U.S. Steel, the only new or "greenfield" (virgin site) steel mill built in the United States during that decade. This suggests the magnitude of the steel industry's investments in iron ore; expenditures for new iron mines and pelletizing plants represented a significant portion (roughly 15 to 20%) of the industry's total investment during the 1950s. In spite of such expenditures, however, trends in world iron ore markets led to the evaporation of the U.S. advantage in iron ore costs.

Viewed from the perspective of the 1980s, the investment performance of the U.S. steel industry during the 1950s seems haunted by disappointing commitments. This was certainly the case for the massive expansion of open hearth capacity and was probably also true for the commitment to North American sources of iron ore. The major investment projects undertaken by the industry in this decade did not serve to substantially increase or even maintain its technological preeminence, even though capacity had been increased enormously and great sums had been spent. As a result, the productivity advantage that defined the cost competitiveness of American steel producers began to erode. In spite of substantial capital expenditures, during the 1950s the U.S. steel industry was marking time.

PROFITS AND PRICES

The threat represented by these trends was only dimly apparent in the 1950s. Weak demand during the latter half of the decade could be ascribed to the disturbing effects of the 1958 recession and the 115-day strike in 1959, while the erosion of the industry's technical advantage did not begin to affect market shares until the end of the decade. Profit rates, the principal indicator by which any industry's performance is judged, were fairly high for the steel industry through much of the 1950s, especially when contrasted with later years (see Table 3–8). In three of the ten years from 1950 to 1959, the rate of return in steel exceeded the manufacturing average, a phenomenon that was to be repeated only once since that decade (in 1974). Even when steel profits fell below the manufacturing average, they were generally close to that level, averaging almost 12 percent a year versus a manufacturing average of slightly more than 13 percent.

Given the trends that we have discussed—especially the divergence between capacity and demand—the overall maintenance of profitability is surprising. As one would expect, profit rates in the industry did fall

Table 2–8. Comparative Price Trends, Steel versus Wholesale Products (average annual rates of increase).

	Steel	*Wholesale Products*
1948–59	5.8%	2.0%
1959–68	0.7%	0.8%
1968–81	9.6%	8.7%

Source: U.S. Department of Labor, Bureau of Labor Statistics.

with operating rates in the latter part of the decade. Yet profitability in the steel industry exceeded the manufacturing average as late as 1957, when the expansion of capacity was largely complete. The key to this puzzle is provided by the data on relative prices presented in Table 2–8.

The extent to which steel prices rose faster than those of other producers' goods during the 1950s is the background to what became one of the most long-standing and bitter controversies concerning steel industry performance in the late 1950s and early 1960s: the prevalence, or absence, of administered pricing in the industry. Oligopolistic industries, it is often claimed, can manipulate prices to maintain profits in the face of weak demand. In the early 1960s, several academic economists accused the steel industry of failing to alter prices with market conditions.[15] The industry consistently denied the charges, and the government in effect provided its imprimatur to the critics through vigorous jawboning against steel price increases—the most celebrated instance of which was the confrontation between President Kennedy and Roger Blough, then chairman of U.S. Steel Corporation.

Given that demand grew hardly at all during the 1950s, particularly in the latter half of the decade, it is highly likely that the persistence of steel price increases was connected to the oligopolistic structure of the industry. This does not mean, however, that excess profits were earned; steel profit rates hovered near the manufacturing average. It is more likely that cost pressures played a major role in steel pricing patterns, especially as rates of capacity utilization fell. Given the substantial investment required for capacity expansion and the fact that expansion far outstripped demand growth in the 1950s, it is clear that price increases were the industry's most attractive option for maintaining profit rates in the face of the rising unit costs caused by low operating rates. Oligopolistic market power was then the condition for price increases rather than its cause. It is ironic that governmental pressure to restrain steel price increases was most intense only ten years after the government had vehemently insisted on increased capacity, while the presence of excess capacity was one of the chief sources of the pressure to increase prices.

This is a much simplified description of the factors involved in the steel pricing controversy, yet it indicates that the combination of forces that acted to boost steel prices was more complex than most parties to the controversy were willing to admit. Regardless of this, however, the maintenance of profit margins through administered price increases is a viable strategy only so long as a concentrated market structure exists. The entry of two new players into the steel market during the 1960s—imports and mini-mills—permanently altered the market structure of the U.S. steel industry. This development restricted the steel industry's ability to vary prices to compensate for discrepancies between capacity and demand, so that profit rates were subject to increasing pressure. Steel price increases lagged the manufacturing average during the 1960s, when average profit rates were actually lower than in the 1970s. Reduced profitability made it difficult for the industry to attract funds for investment, and this in turn made it impossible to maintain the industry's technical leadership and cost competitiveness. By the end of the decade, the vicious circle had begun to spin.

CONCLUSION

1959 was a watershed for the American steel industry. The year was dominated by a long and bitter strike, the effects of which are still felt. By that year the robust expansion of capacity was complete, even though it was clear by the end of the 1950s that the basic oxygen furnace was commercially viable and that the industry would have to undertake the transition to that technology. Most importantly, steel imports exceeded exports in 1959 for the first time in the twentieth century. While imports are not necessarily a cause of the industry's decline, they are certainly one of its most public manifestations. The dormancy of the import problem until 1959 is one of the most persuasive indicators that in this chapter we have been dealing with an era different from the present one. Nevertheless, the trends that determined the industry's subsequent decline emerged during the 1950s. These trends have drastically transformed the American steel market, and changes in the U.S. steel industry have failed to keep pace.

Overall, the sluggish growth in steel demand during the 1950s is the most salient feature of that decade. Clearly, there had been a serious misreading of the trends in steel demand in the late 1940s and early 1950s. It is little solace for the industry to recall that the government had erred more wildly in its projections of demand growth, since it was the industry that directly sufferred the consequences. The most obvious consequence was the steady erosion of operating rates, as new capacity came onstream while demand remained essentially flat. Because of its failure to accurately forecast trends in consumption, the industry found itself in a "scissors crisis" described by the relationship between capacity expansion and stagnating demand.

The most significant consequence of the industry's inaccurate demand forecasts was the commitment to substantial expansion at a time when the technology of steelmaking was about to be revolutionized. Even robust growth in demand would not have saved the industry's open hearths from technological obsolescence by the mid-1960s. Had steel capacity grown only with demand, however, the inevitable transition to the basic oxygen furnace would have been eased. A more sober, conservative, and realistic investment strategy would have saved the industry much of the anguish eventually connected with the abandonment of the open hearth.

The persistent inflation in steel prices during the 1950s can also be ascribed to the failure of demand to grow at a rate approaching the pace of capacity expansion. This point is more open to debate; it is possible that increased steel prices themselves acted to suppress demand and to encourage the use of substitute materials. Certainly the industry pursued its pricing policy under the assumption that demand was highly inelastic. Regardless of this, steel industry profits in the 1950s were not exorbitant when compared with returns in other industries; hence, there is a basis for arguing that the price increases served to maintain profit margins in the face of rising costs as capacity utilization fell. It seems plausible that had demand grown with capacity—that is, had the industry's demand forecasts been accurate—steel price increases may well have been more restrained. The potential impact of lower steel prices on import penetration and eventual changes in market structure is obvious, although it must remain a matter of conjecture.

The fundamental consequence of the anemic growth in steel consumption was the restriction of the industry's room to maneuver and the elimination of much of the managerial margin for error. Given the increasingly complex steel market of the 1950s— especially the quickening tempo of technical change and the fundamental shift denoted by the emergence of a truly international steel market—the effects of mistaken investment decisions could very quickly turn ominous. This danger was especially great without the recuperative impact of rapid market growth. Unlike its major competitors, the U.S. steel industry did not enjoy such growth. Yet it acted as though its preeminence was unassailable: investment was undertaken with little regard for the potential of technological progress, and oligopolistic behavior was sustained in spite of increasing evidence of potential entry. Hence, in spite of its still formidable size and technology, the position of the U.S. steel industry in 1960 was drastically different from what it had been in 1950. Its dominance in scale and technology no longer seemed invulnerable, recent investments had failed to boost efficiency, and its market was growing much more slowly than those in Europe and Japan.

Notes

1. *The New York Times*, January 2, 1951, p. 67.
2. Joseph S. Bain, *Barriers to New Competition* (Cambridge, Mass.: Harvard University Press, 1956), p. 236.
3. Anthony Cockerill, *The Steel Industry; International Comparisons of Industrial Structure and Performance* (Cambridge: Cambridge University Press, 1974), pp. 39–66.
4. Economic Commission for Europe, United Nations, *Long-term Trends and Problems of the European Steel Industry* (Geneva: United Nations, 1959), pp. 80–113.
5. American Iron and Steel Institute (AISI), *Annual Statistical Report, 1960* (New York: AISI, 1961), p. 9.
6. Economic Commission for Europe, *Long-term Trends and Problems of the European Steel Industry*, p. 1.
7. Donald F. Barnett, *Economic Papers on the American Steel Industry* (Washington, D.C.: AISI, 1981), p. B–6.
8. AISI, *Iron and Steel Works Directory of the United States and Canada* (New York: AISI, 1951 and 1960), pp. 482 and 448, respectively.
9. Ibid., pp. 476 and 442, respectively.
10. Ibid,, pp. 483 and 449, respectively.
11. "Why Steelmen Raise the Ante," *Business Week*, November 16, 1963, pp. 144–46.
12. Much of the data presented in the discussion of this theme are presented in William Hogan, S.J., *History of the Iron and Steel Industry in the United States* (Lexington, Mass.: D.C. Heath, 1972), pp. 1481–96.
13. Calculated from consumption and import data in various sources, e.g., Organization for Economic Cooperation and Development (OECD) and U.N. reports, AISI *Annual Statistical Reports* (New York and Washington, D.C., various years), "Tex Reports" (Tokyo, various years).

14. AISI, *Annual Statistical Report* (Washington, D.C.: AISI, various years).
15. See, for example, Walter Adams and Joel Dirlam, "Steel Imports and Vertical Oligopoly Power," *American Economic Review* 54 (June, 1964): 626–55.

3 THE END OF AN ERA, 1960–1982

We have been shocked out of our complacency and smugness. We now realize that American industry has no manifest destiny to be always first, always right, always best.[1]

—David Roderick
Chairman, U.S. Steel Corporation
May 1982

The principal strategic error of the American steel industry in the 1950s involved the failure, given the slow growth in its home market, to target its investments toward performance improvements. This failure was not evident in that decade; financial performance was relatively good until the recession of 1958. Yet the industry had lost momentum relative to its international rivals, and by the 1960s managerial options were constrained by the legacies of excess capacity, a misguided commitment to the open hearth, and the problematic reliance on North American sources of iron ore.

During the 1960s the industry mounted a major effort to maintain its competitive status; investment, chiefly devoted to replacing the open hearth with the basic oxygen furnace, reached its highest postwar levels during that decade. Yet this effort did not generate rapid enough improvements in performance. The industry still lost ground relative to foreign competitors, and profitability was generally far below the manufacturing average. By the 1970s the U.S. industry had lost its competitive advantage and, more importantly, had no viable strategy for reversing its decline. The results have been persistent loss of market share, intense efforts to stem the flow of imports via trade barriers, and deteriorating profitability. These trends culminated in the disastrous year of 1982, when the industry lost over $3 billion.

This seemingly inexorable decline can be attributed to several factors. Most fundamentally, however, it is due to far-reaching changes in the American and world steel markets. In the United States, the industry has been confronted with the slow growth described in Chapter 2. Rather than develop a strategy attuned to slow growth, however, the industry has clung to an optimistic faith that periods of weak demand will be transitory. In a world context, the U.S. industry has been victimized by the gradual shift of production away from Europe and North America, the traditional steel-producing centers. Here again, the U.S. industry has sought to avoid the implications of this trend, in this case by seeking protection from the government. On both fronts, U.S. producers have failed to develop realistic adjustment strategies, so that the original American advantage has been eroded by the more favorable conditions, particularly market growth, enjoyed by foreign competitors.

This chapter will begin by discussing the reasons for the weak growth in U.S. steel consumption and the changing pattern of world steel production. Together, these define the principal features of the economic environment faced by American steel producers. The pressures they have engendered have led to lagging investment and a quickening deterioration in the technological position of the U.S. industry. This in turn has placed U.S. producers at an increasing disadvantage in terms of costs, so that profits have been squeezed. These are the themes with which this chapter will close: investment, technological performance, cost trends, and profitability. The internal structural changes through which the U.S. steel industry has responded to these developments will be discussed in Chapter 4.

CHANGES IN MARKETS

The aggregate data presented in Chapter 2 (Table 2–7 and Figure 2–2) showed that from 1950 to 1981 U.S. steel consumption grew at an anemic annual rate of 1 percent. In spite of this trend, the industry has traditionally been fairly optimistic about the prospects for steel demand. Periods of peak demand (such as 1973–74) have tended to vindicate the more optimistic forecasts, while periods of weak demand are perceived as the normal cyclical compliment to the peaks or as the consequence of transitory, anomalous circumstances. This optimistic bias, while possibly gratifying to management, has clouded the industry's perception of the underlying forces that determine the secular trend in steel demand.

The most general relationship defining the trend in steel consumption is referred to as steel intensity; it relates steel's share of GNP to the overall standard of living (per capita GNP). Declining steel intensity is characteristic of maturing economies, even though the absolute level of steel intensity varies according to the structural characteristics of the

Figure 3–1. Steel Intensity in Industrialized Countries; Actual Fluctuations and Average Curve.

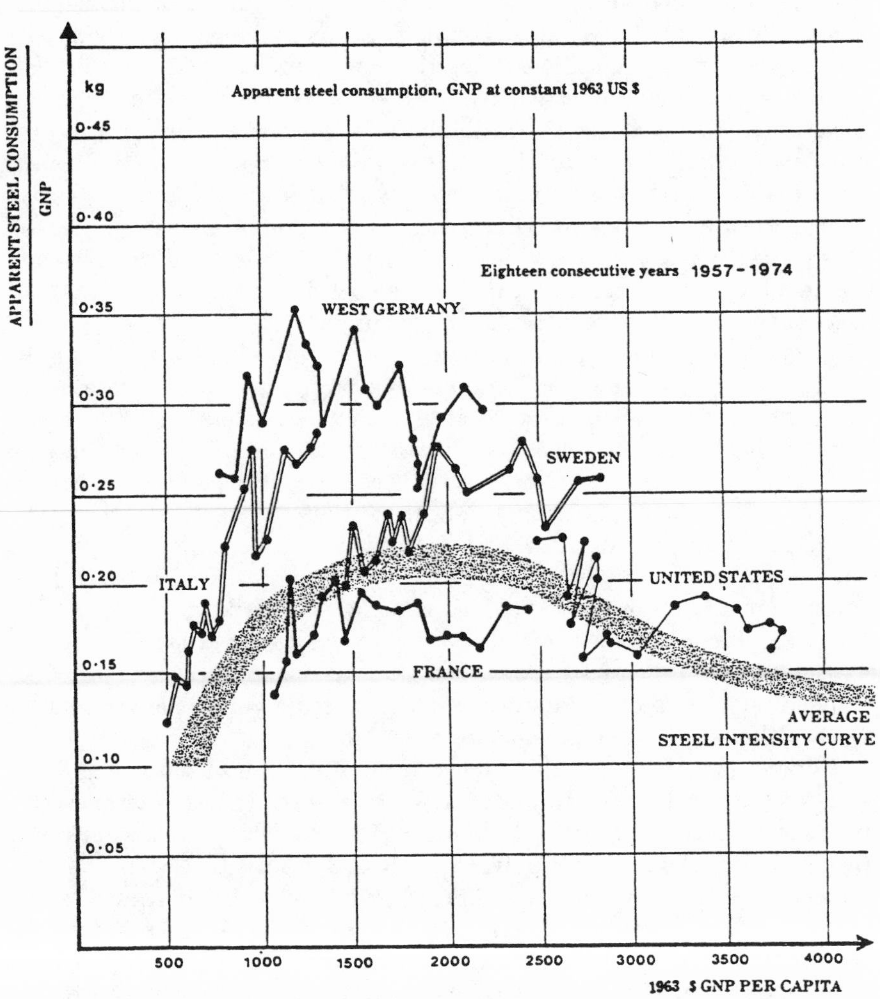

Source: International Iron and Steel Institute (IISI), *Steel Intensity and GNP Structure* (Brussels: IISI, 1974).

country concerned. This is shown in Figure 3–1, which actually understates the decline in steel intensity, since its data series ends with the energy crisis and the steel boom of 1973–74. Several arguments are traditionally adduced to explain the pattern described in Figure 3–1. The most basic is the plausible premise that the early stages of industrialization require relatively steel-intensive investments in infrastructure (roads, bridges, railroads, port facilities, power stations, etc.), which no longer predominate once the industrialization process is advanced.

A second explanation for declining steel intensity is the fact that increased technological sophistication implies the development of new materials (or whole new industries) that compete with steel, the traditional basic material of an industrialized society. This in turn implies a relative contraction in the uses for which steel is the preferred material. Steelmakers have responded to this challenge by developing thinner and lighter steels, but success on this front also reduces steel intensity, since it limits the tonnage required for a specific application.

The best examples of products that have gained markets at the expense of steel are plastics and aluminum. Competition with aluminum in the container market has forced steel suppliers to develop thinner and more malleable tinplate products–products that in effect have the look and feel of aluminum. Nonetheless (possibly because of consumer tastes), aluminum now dominates the market for beverage cans. The aluminum share of this market increased at a spectacular rate during the 1970s: from 22.6 percent in 1972 to 84.7 percent in 1981 (steel accounts for the residual).[2] Steel has retained its position in the market for food cans, but this sector offers few prospects for growth, especially in the face of potential innovations such as plastic pouches. In 1981, steel shipments to the container industry were 24 percent lower than in 1971, although total shipments were unchanged.[3]

Competition between steel and aluminum, while intense, has been sedate compared to the inroads made by plastics in markets that had been dominated by more traditional materials. With the exception of products for which strength and weight are desirable (e.g., plates and beams), plastics have gained markets at the expense of almost every category of steel product. Even after the energy crisis, when their relative price ceased to decline, plastics continued to make inroads in markets where qualities like weight reduction, formability, and corrosion resistance are attractive properties. This is particularly apparent in the automobile industry. The Ford Motor Company, for instance, reduced the weight of its cars by 28 percent from 1977 to 1982. Steel content, by weight, dropped 30 percent (hot- and cold-rolled steel content fell 39%), while plastic content rose 36 percent by weight. Whereas Ford's cars contained fourteen times as much steel as plastic in

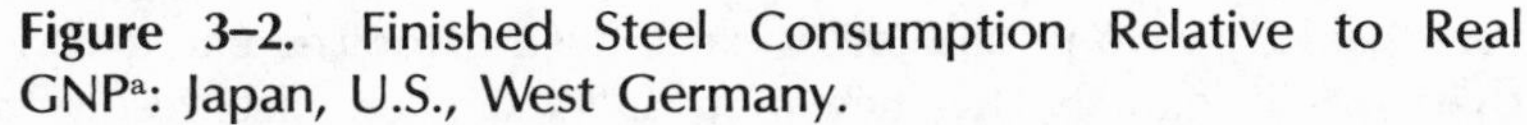

Figure 3–2. Finished Steel Consumption Relative to Real GNP[a]: Japan, U.S., West Germany.

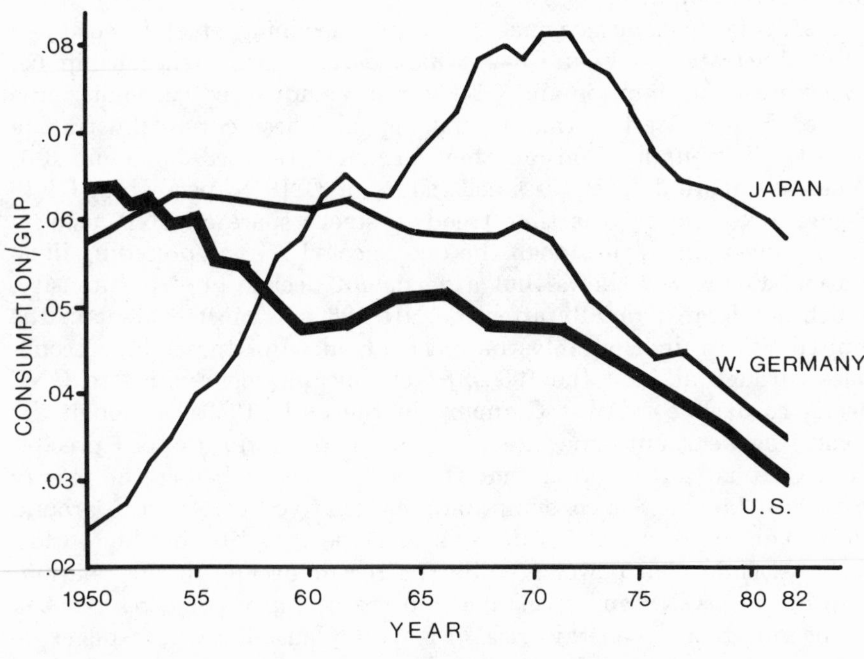

a. Tons per thousand units of respective 1981 currency, in 1981 U.S. dollar equivalents. The ratios for each country were calculated using respective currencies (1981 terms) then, for ease of illustration, were converted to U.S. dollars using 1981 exchange rates. Comparative levels of steel consumption relative to GNP are not meaningful; only changes over time are relevant.

Sources: Calculated from International Monetary Fund (IMF), *International Financial Statistics* (Washington, D.C.: IMF, various years); American Iron and Steel Institute (AISI), *Annual Statistical Report* (Washington, D.C.: AISI, various years); International Iron and Steel Institute (IISI), *Steel Statistical Yearbook* (Brussels: IISI, various years); World Steel Dynamics, *Core Report R* (New York: Paine Webber Mitchell Hutchins, 1982); and authors' estimates of raw steel to finished product yields.

1977, this ratio had been cut in half by 1982.[4] In 1981, steel shipments to the automobile industry were 25 percent lower than in 1971.[5]

It is doubtful that steel will be replaced as the fundamental material for industrial societies. Nevertheless, plastics and aluminum are likely to continue to make gains in markets that have traditionally consumed large amounts of steel. While steel intensity has been declining, the equivalent ratios for aluminum and plastics have been increasing—at a very healthy rate in the latter case, even after the energy crisis.[6] Furthermore, it is likely that new materials (e.g., ceramics, resin composites, metallic powders, etc.) will emerge to limit the markets for which steel is the preferred input.

Maturing economies typically exhibit declining steel intensity–a point illustrated in Figure 3–2, which describes the relationship between steel consumption and GNP for three advanced economies: the United States, West Germany, and Japan. These curves illustrate a slightly different relationship than the one described in Figure 3–1. Whereas Figure 3–1 relates steel's share of GNP to per capita GNP, Figure 3–2 describes the time trend in steel's share of GNP, using a much longer time frame than the one selected for the preceding illustration. Figure 3–2 shows that a persistent decline in steel intensity, which accelerated rapidly after the late 1960s, is most evident in the United States–indisputably the most advanced of these three economies through at least the 1960s. Steel consumption relative to GNP began to decline in West Germany in the early 1950s, although the trend was weak until the late 1960s, when the ratio dropped precipitately. (It is worth noting that the decline began before the energy crisis.) In Japan, steel consumption grew relative to GNP until around 1970. This pattern reflects Japan's emergence as one of the world's foremost industrial powers. While the recent decline in the relationship between GNP and steel consumption in Japan could be ascribed to the effects of the energy crisis, it is more plausible to view this experience as the onset of the type of pattern that has been prevalent in the United States since World War II and in West Germany since the late 1960s. All of this evidence suggests that the inertial trend in U.S. steel consumption favors slow growth, that is, growth rates lower than those achieved in the economy as a whole.

Declining steel intensity is a secular tendency that counteracting circumstances can occasionally offset. This was the case during the 1960s, for instance, when the ratio of steel consumption (millions of tons) to GNP (billions of 1981 dollars) rose from 0.049 in 1961 to 0.051 in 1968; this was after a sharp decline, from 0.064 in 1950, during the previous decade (see Figure 3–2). Steel executives' hopes for a resurgence of strong demand–in other words, a denial of the inevitability of declining steel intensity–are based on the experience of this period as well as on the steel boom of 1973–74.

Closer investigation suggests that the steel boom of the mid-1960s was an exception linked to the strong overall growth of the U.S. economy during that period. This prosperity was due in large part to the application of Keynesian principles to federal economic policy. In an effort to push the economy toward its GNP potential, tax policies were enacted to boost capital investment, an aggregate that is relatively steel intensive. Furthermore, government expenditure on infrastructure, especially the interstate highway system, reached its peak in these years. Such expenditures are also steel intensive.

By the 1970s, these offsetting factors were no longer operative, so that the underlying trend in steel intensity reasserted itself. The ratio of steel consumption (millions of tons) to GNP (billions of 1981 dollars) fell from 0.048 in 1970 to 0.032 in 1981—an annual decrease of 4 percent. The same rate of decline in this ratio was evident in West Germany during this decade, while steel consumption relative to GNP fell 3 percent per year in Japan. In the United States, steel demand has been highly volatile during the 1970s; the decade included both the strongest boom (in 1973–74) and the deepest bust (in 1975) of the entire period from 1950 to 1980. Although this volatility has persisted since 1975, the years of weak demand have been more frequent than have the compensating good years. Weak markets culminated in the crushing collapse of 1982, when steel shipments fell 31 percent from the middling level of 1981.

Economic recovery will not reverse the declining trend in steel intensity. The secular forces that reduce steel intensity (e.g., competing materials and the reduced importance of infrastructural investment) are now being supplemented by profound structural changes in the American economy. These changes, associated chiefly with the development of advanced electronics, are altering long-standing patterns of production and consumption in ways that are reducing the relative importance of steel. Tastes are shifting away from traditional consumer durables and toward the new products that are being generated by advanced electronics. Business expenditures are increasingly being devoted to investments (e.g., data processing) that are much less steel intensive than traditional investments. Such shifts have been gestating within the U.S. economy for well over a decade, but their pace has been quickened by the energy crisis and by the austerity enforced by stagflation and recession. An accelerated decline in steel intensity is one aspect of these changes, which are revising the standards of competitiveness for every industry.

CHANGES IN COMPETITION: THE "IMPORT PROBLEM"

Declining steel intensity has been but one of the problems faced by the U.S. steel industry, since its shipments of steel products have failed to keep pace even with the laconic growth of consumption. Imported steel has captured most of the expansion in the U.S. steel market, a situation that reflects basic structural changes occurring within the world steel market.

Shifts in the Global Pattern of Production and Trade

Since World War II, significant changes have occurred in how and where steel is produced and consumed internationally. The most profound change has been a basic shift away from the traditional steel centers in Europe and the United States; the dominance of these regions, in both consumption and production, has weakened significantly since the 1950s. Second, steel markets have become internationalized to a far greater degree. The proportion of steel production involved in world trade grew from 10.7 percent in 1950 to 25 percent in 1980.[7] Interregional steel trade–excluding trade within the European Economic Community (EEC) and within the Council for Mutual Economic Assistance (COMECON)–has grown somewhat more slowly, but it has still increased at a compound annual rate of over 7 percent during the past thirty years. By contrast, over the same period total world production grew 4.7 percent per year and Western world steel production by 4.5 percent. More recently, these structural changes have been accompanied by the ominous development of substantial excess capacity–a situation that has become painfully apparent since 1975 and that is likely to accelerate the structural changes occurring within the world steel industry.

Both the international diffusion of steelmaking and the development of excess capacity are illustrated in Figure 3–3, which refers only to the period after 1960. Were one to look further back, the relative decline in the role played by European and American steel producers would be even more pronounced. In 1937, for instance, 92 percent of the world's total steel output (excluding the Soviet Union) was produced in Europe or North America.[8] By 1961, those regions' share of Western steel output had fallen to 82 percent; in 1980, it was 60 percent.[9] Even if industrialized countries outside Europe and North America–especially Japan, the most stunning manifestation of the shift away from Europe and the United States–are excluded, the diffusion of steel production and consumption is still unmistakable. In 1961, the developing countries' share of world steel production and consumption were 4.1 percent and 9.6 percent respectively; by 1980, these figures had increased to 12.4 percent and 21.2 percent.[10] Finally, steel production and consumption has grown most rapidly and consistently in nonmarket economies. Since these countries do not participate in the Western steel market to any great

Figure 3–3. Western World Steel Capacity and Production[a], 1960–1982[b] (disaggregated by major region, 3-year moving average).

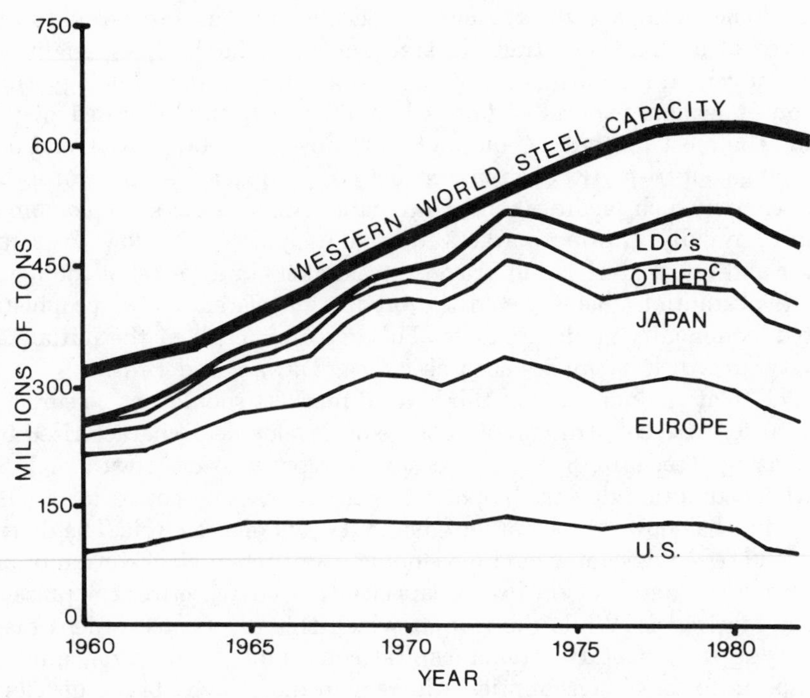

a. In millions of tons of raw steel.

b. Estimates used for 1983.

c. Other industrialized countries (Canada, South Africa, Australia, New Zealand).

Sources: World Steel Dynamics, *Core Report R* (New York, Paine Webber Mitchell Hutchins, 1982); Organization for Economic Cooperation and Development (OECD), *The Steel Market in 1981 and the Outlook for 1982* (Paris: OECD, 1982); and International Iron and Steel Institute (IISI), *Steel Statistical Yearbook* (Brussels: IISI, various years).

extent (although even this is changing slowly), they will be excluded from the analysis developed below. Nevertheless, their share of total world steel production has grown from 18.6 percent in 1950 to 35 percent in 1981.[11] Their crude steel output has grown at an annual rate of 7.6 percent over that period; by the early 1970s, the Soviet Union had replaced the United States as the world's largest producer of raw steel.

Increasing steel consumption in industrializing regions has encouraged the installation of new capacity in such regions, and the postwar growth of the Japanese economy has provided them with a development model in which steel plays a prominent role. Steel projects are often given high priority in the economic planning of newly industrializing countries, while steel imports are often kept out of such markets once domestic capacity is in place. These trends have contributed to the growth of an extensive trade in steelmaking technology, especially as more advanced industries—e.g., Japan's—have turned to technology sales to offset lagging exports of finished products. Optimally scaled plants using the most up-to-date equipment are now being built in several developing countries. Although infrastructural inadequacies can reduce the efficiency of such facilities, under favorable circumstances the combination of rapidly growing markets, up-to-date plants, and low costs for labor and raw materials can provide steelmakers in developing countries with substantial advantages in the production of relatively unsophisticated, commodity-grade products. This explains much of the initial capacity growth in countries such as Korea, Taiwan, and Brazil.

The relative dynamism of third world markets should not be surprising, since these countries are at a stage of development characterized by increasing steel intensity. In a sense, the slow rate of growth in U.S. steel consumption during the postwar period was a precursor for a pattern that has now gripped other advanced economies as well. The divergence between developed and developing countries in steel consumption trends is one source of the excess capacity that characterizes the present world steel crisis. While the rate at which this divergence widens may slow due to budgetary strains, capital constraints, and so on, growth prospects for steel consumption will remain more favorable in developing countries.

The problem of excess capacity for the world steel industry is immediately apparent in Figure 3–3. Whereas capacity and consumption in the Western world maintained a close relationship prior to the mid-1970s, there has been a sharp break in the trend of consumption since that time. This drop in demand could be ascribed to the energy crisis and its aftermath, yet it would not have been so precipitous were it not for the downward secular trend in steel intensity in the developed countries. Since such weakness in demand was not widely anticipated, steel capacity continued to grow at the pre-1975 pace.

The principal reasons for the overly optimistic expansion of world steel capacity were governmental intervention and increased competition for export markets. By the late 1960s, investment in the world steel industry had largely ceased to be determined by economic criteria—especially in Europe and in some developing countries. In the latter case, steel projects were funded to promote industrialization and import substitution. In Europe, governmental funding of modernization and expansion occurred within the context of policies designed to maintain employment or to foster regional development. The momentum for such projects had little to do with realistic forecasts of future consumption. Similar failures of foresight were linked to the common strategy of building massive plants oriented toward export sales. For developed countries, exports offered the prospect of ameliorating declining steel intensity in domestic markets. For developing countries, exports can offset the limitations of small domestic markets and provide needed foreign exchange.

Increased competition over export markets has encouraged many producers to seek the maximum economies of scale by building well ahead of the market. Governments have played a significant role in encouraging and funding this development. With the exception of the United States, all major producing countries boosted capacity far more than was justified by the growth in consumption during the 1970s. In Belgium and Luxemburg, for instance, capacity grew at an annual rate of 3 percent from 1970 to 1980, while consumption *fell* at an annual rate of 3.4 percent. In Germany, the comparable figures were 3.3 percent and -1.3 percent; in France, 2.5 percent and -1.3 percent; in Japan, 6.8 percent and 1.2 percent.[12] The resultant crisis of low operating rates and intense price competition has led to further governmental involvement in the form of subsidies for operating losses.

The present crisis in the world steel industry is a competitive struggle over where capacity reductions will occur. This struggle, although subject to substantial political interference, will be resolved in favor of the low-cost producers. Outmoded facilities in the United States and Europe are the most likely candidates for closure, so that the present crisis is likely to accelerate the structural shift away from traditional steel-producing regions. The U.S. industry may with some validity protest that it has not contributed to the problem of global overexpansion, but the competitive pressure provided by imports is likely to ensure that it contributes to the solution.

Imports in the American Market

It would be hard to imagine a reader who does not have some impression of the steel industry's import problem. Certainly, this is a theme that has been repeatedly stressed–if not bludgeoned–by the industry. Frustrated by low operating rates and limited growth, American steel firms have been outraged at the persistent penetration of their markets by foreign suppliers. During the 1950s, the deteriorating competitiveness of the U.S. steel industry was evident in the gradual elimination of the

Figure 3–4. Import Share of Integrated Markets and Total Market, 1955–1982.

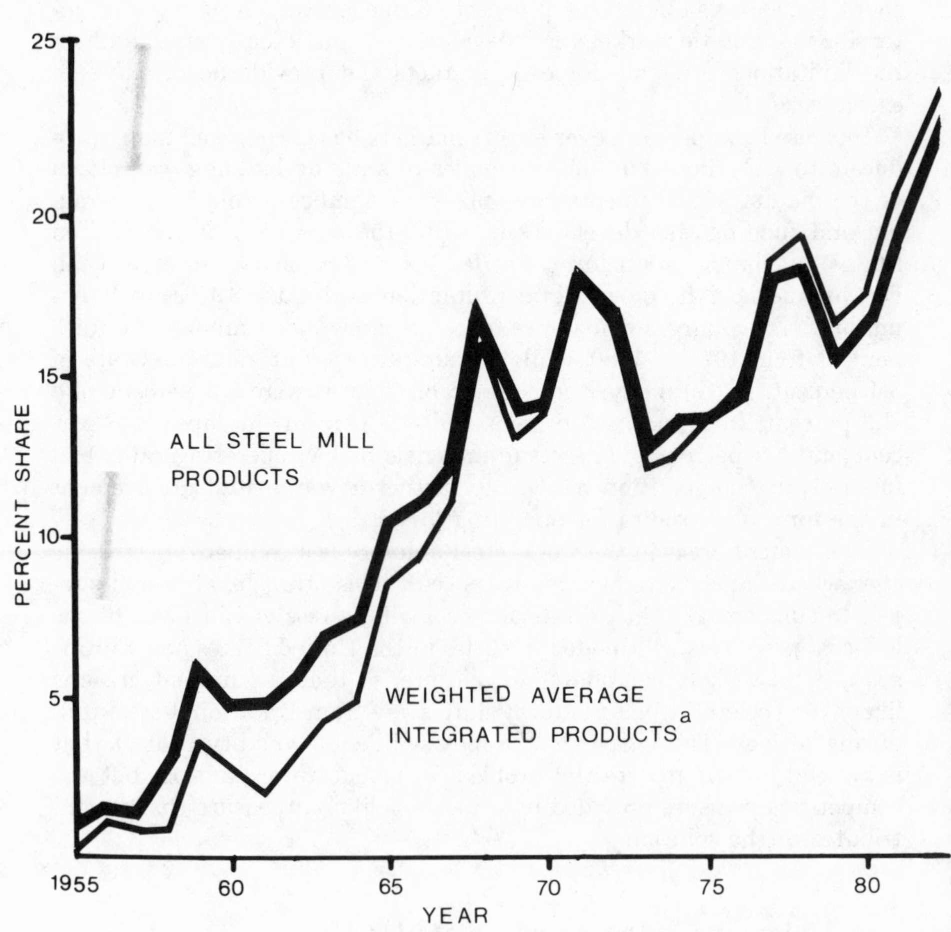

a. Integrated composite, weighted by total shipments, composed of flat-rolled products (hot- and cold-rolled sheet, galvanized, tinplate), plate, heavy structurals, and pipe and tube.

Note: Contract negotiations associated with potential strikes occurred in 1959, 1962, 1965, 1968, and 1971.

Source: American Iron and Steel Institute (AISI), *Annual Statistical Report* (New York and Washington, D.C.: AISI, various years).

Table 3–1. U.S. Imports of Steel Mill Products: Shares by Source[a] (%).

	Europe	*Japan*	*Canada*	*Other*
1961	66.6	21.4	8.3	3.7
1971	47.2	39.1	7.2	6.5
1981	37.2	33.9	13.8	15.1

a. Calculated from three-year average around year listed.
Source: American Iron and Steel Institute (AISI), *Annual Statistical Report* (New York and Washington, D.C.: AISI, various years).

industry's status as a net exporter. In the early 1960s, steel imports gained a significant share of the U.S. market for low-value commodity products—a market that was somewhat peripheral to the operations of the major U.S. companies. By the mid-1970s, however, imports had gained substantial shares in almost every product line (with the peculiar exception of some low-value products; see Chapter 4). Even the industry's core market in the Midwest and the crucial automobile sector were no longer sacrosanct. Figure 3–4 describes the overall trend in steel imports and shows that since 1975 imports have increased their share in integrated markets more rapidly than in the U.S. market as a whole.

The sources of U.S. imports have shifted in accordance with the overall trends in world steel production. Just as nontraditional steel-producing countries have emerged as a major force in the world market, they have gained an increasing presence in the U.S. market as well. Table 3–1 describes the share of total steel imports from various sources over the past two decades, when steel imports have grown from 4.7 percent of the domestic market to 19.7 percent. Since 1961, there has been a drastic decline in the relative importance of steel imports from Europe. The principal beneficiaries of this decline have been Japan and other nontraditional steelmaking centers (e.g., Latin American, Korea, South Africa). The increasing weight of the "other" category, which is more or less synonymous with developing countries, is likely to continue, as is the declining significance of imports from Europe.

From a global perspective, increased imports reflect a profound shift in the pattern of production; from the domestic perspective, they represent one form of entry into what had been a highly concentrated, oligopolistic industry. When imports are combined with mini-mill shipments, the market share of the traditional integrated sector of the American steel industry has fallen from about 95 percent in 1960 to about 60 percent in 1982. The emergence of import competition has exacerbated the effects of declining steel intensity, and it has contributed to a redefinition of competitive standards within the U.S. steel market. Prior to 1959 the American steel market was basically autarchic and was dominated by the traditional firms; this situation has since disappeared. From 1960 to 1981, U.S. shipments grew at an annual rate of 1 percent, while consumption increased 1.8 percent per year. Imports have thus captured almost half the growth in the U.S. market. This problem has grown worse for the U.S. industry since the late 1960s, since demand has failed to grow significantly while imports have increased.

In the industry's view, the most pernicious effect of steel imports is the restraining effect they have on profitability, prices, and investment. Since steel is a capital-intensive industry, incremental fluctuations in output have significant consequences for costs. Even small increases in imports thus have disproportionately large effects on profits. Furthermore, domestic firms lose sales if they do not meet import prices; this also reduces the profit margins realized by American steelmakers. This loss of pricing power is probably one of the more bitter pills for steelmakers to swallow, given the oligopolistic market power they enjoyed prior to 1960. Finally, since increased import penetration has reduced the domestic industry's share of consumption growth, the attractiveness of investments has been reduced. Replacement investment is always harder to justify than are investments that also boost capacity.

While these incremental effects of increased imports are undoubtedly important, their most significant effect has been to alter the standards of competitiveness within the U.S. market. The more favorable conditions enjoyed by foreign producers, especially in terms of market growth, have provided them with a significant competitive advantage. This has enabled foreign producers to gain a technological advantage vis-a-vis their U.S. competitors.

American producers have attempted to cope with the onslaught of imports in several ways: for example, increased investment during the mid-1960s and contractual innovations with the United Steelworkers, both of which are discussed later in this chapter. Their chief response, however, has been an increasingly intense campaign for some form of trade relief. The industry has succeeded in winning several programs from the U.S. government, and it has constructed a rather potent lobbying force by combining with the United Steelworkers on the trade issue. The political clout of this coalition led to the formation of the Congressional Steel Caucus in the mid-1970s; this group, the defender of the domestic steel industry, has a membership of some 180 members of Congress. Unfortunately for the industry, however, its lobbying efforts have had only marginal and transitory effects on the trend in imports. Steel trade policy has failed to address the structural causes of the domestic industry's declining competitiveness, of which increased imports are only a manifestation. Hence, as each trade program fades away, the problem reemerges with greater force.

Most recently, the industry has focused its anti-import campaign on the prevalence of governmental subsidies, especially in Europe. Such subsidies undoubtedly exist; state funding of steel operations in Europe—particularly Britain, France, Italy, and Belgium—has seriously distorted price-cost relationships in international steel trade. State involvement has encouraged the uneconomic construction of excess capacity, and the low operating rates that this has provoked have necessitated massive transfusions of governmental funds. Dependent on private capital markets, American producers have found it increasingly difficult to compete with subsidized operations. In spite of the undeniable presence of unfairly traded imports in the U.S. market, however, the U.S. steel industry has been arguing an increasingly irrelevant case. It has clearly lost its competitive edge against the Japanese and is falling behind other steel industries as well. While the industry's arguments on the import question have changed little since the emergence of the import problem, its position in the marketplace has deteriorated significantly.

THE END OF TECHNOLOGICAL SUPREMACY

Lagging growth in demand and the emergence of new competition have defined the American steel industry's principal strategic problems during the postwar period. Through the 1950s, the cost competitiveness of the American steel industry was based on significant advantages in scale and technology. Together with the relatively strong resource base of the U.S. industry, this technical advantage was sufficient to offset an extreme disadvantage in employment costs. Hence the fundamental challenge confronting U.S. steel firms was how to maintain their technological competitiveness in a low growth environment. This is a challenge that by and large has not been met.

Fueled by rapid growth in their home markets and by a burgeoning trade in steel products, some foreign producers were able to eliminate the American technological advantage. Fortuitously, their major expansion programs coincided with the commercialization of the basic oxygen furnace, which had become the preferred steelmaking technology by the late 1950s. Expanding foreign producers, whose markets rapidly approached that of the U.S. industry in terms of size, eventually equaled or surpassed their American competition and in the process redefined the standards of technological excellence, which had previously been set

Figure 3–5. Steel Industry Investment[a] Relative to Capacity, 1950–1981[b] (1981 dollars per net ton, 5-year moving averages).

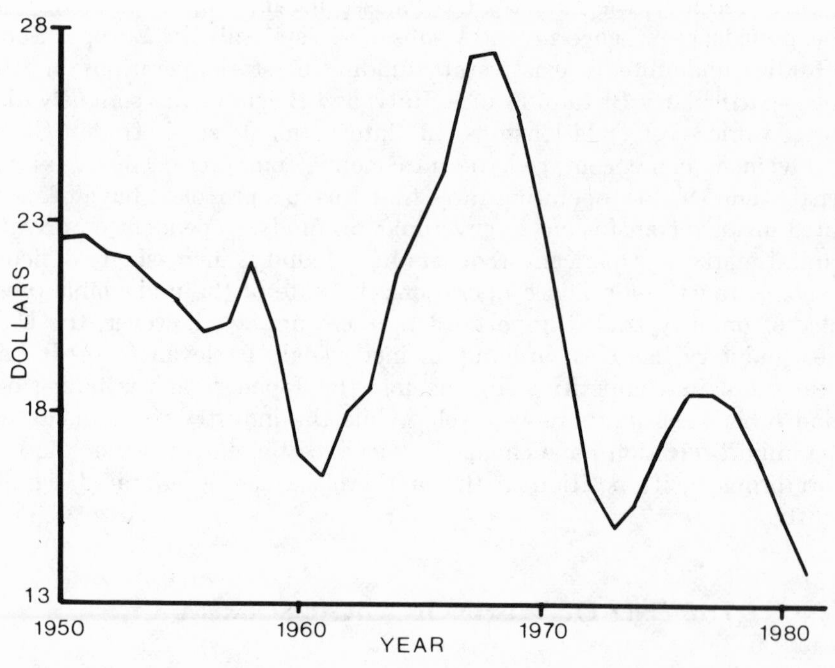

a. Productive investment (environmental and nonsteel expenditures excluded).
b. Estimates used for 1983.
Sources: American Iron and Steel Institute (AISI), *Annual Statistical Report* (New York and Washington, D.C.: AISI, various years); Council on Wage and Price Stability (COWPS), *Prices and Costs in the U.S. Steel Industry* (Washington, D.C.: U.S. Government Printing Office, 1977).

by the United States. This was especially true of Japan, which provides an outstanding example of how the proper strategy can transform competitive weaknesses into strengths.[13]

Lagging Investment

Although constrained by relatively weak market growth and deteriorating profitability, by the 1960s American steel firms were faced with the need to mount major investment campaigns in order to maintain competitiveness. Productive capital expenditures in the steel industry rose during the 1960s, averaging $3.4 billion (in 1981 dollars) for the decade as a whole and reaching an all-time high by the middle of the decade (see Figure 3–5). This annual average represented an increase of over 25 percent relative to the performance of the previous decade, in spite of the fact that capacity grew little during the 1960s. The earlier concentration on capital-widening investment gave way to a capital-deepening focus. Whereas the industry had made little technological progress during the preceding decade, investment during the 1960s was devoted to massive modernization programs, particularly the adoption of the basic oxygen furnace (BOF). A secondary target of investment was the installation of the continuous hot strip mill—an innovation that greatly increased the speed and scale of such facilities and which in a sense represented the steel industry's adoption of the continuous assembly techniques characteristic of the major industrial changes from 1870 to 1920. Modernizing investments like these made significant contributions to productivity and efficiency, unlike the capital-widening investments of the previous decade.

Although both the basic oxygen furnace and the continuous hot strip mill were first installed in the 1950s, rapid adoption of these technologies did not occur until the following decade. By that time internal cash flow was inadequate to support the targeted level of expenditures; as a result, steel firms turned to capital markets to sustain the increase in expenditures described in Figure 3–5. The average ratio of long-term debt to equity rose from 24 percent in 1960 to 38 percent in 1970, and to over 50 percent in 1980.[14] While this increase in debt enabled the industry to undertake the transition to the BOF, it is clearly a financial tactic that has an inherent limit. By the 1970s, this limit had been reached. Further significant increases in the debt-to-equity ratio were no longer feasible, and new equity issues were no longer attractive. As a result, the industry's investment fell drastically.

Figure 3–5 shows the extent to which steel industry investment has fallen since the late 1960s. Investment during the 1970s and 1980s has clearly been too low to sustain the industry's technological advantage—a point that will be discussed in depth in Chapter 6. Total productive investment averaged only $2.6 billion a year (in 1981 dollars) during the 1970s—more than a 25 percent drop from the level reached in the previous decade. As a result, it has become increasingly difficult for the industry to execute capital-deepening programs like the adoption of the BOF.

The adoption of the BOF represented the industry's last gasp in terms of major investment programs. Since this effort failed to reverse or even slow the industry's competitive decline, the return on investment was meager, discouraging similar efforts subsequently. Although American producers in the 1960s attempted to match the investment efforts of their foreign competitors, slower market growth ensured that the U.S. performance would be relatively patchy and incomplete. Major successes like the adoption of the BOF masked deteriorating technical standards in other processes.

The problems associated with a generally low rate of investment were compounded by the fact that postwar investment booms in the steel industry have often had a lemminglike character. The same market opportunities are sniffed out by several firms, each of which moves to add capacity or to modernize facilities. The net result for the industry as a whole is the effective dilution of the market opportunity. Excess capacity restrains prices and thus lowers the return on investment. This pattern is especially problematic at the finishing end, and it has developed in several categories of facilities: continuous hot strip mills during the 1960s, wire rod mills during the early 1970s, and seamless tube mills during the early 1980s. This suboptimal pattern of investment for the industry as a whole stems in part from steel firms' lack of specialization. If all firms seek to provide a more or less complete line of products, modernizing investment by one firm must be matched by all others. Since modernization generally entails increased capacity, the result is excess capacity in that product line for the industry as a whole. Such a pattern is generally avoided in other countries (e.g., Canada and Japan), where the government often encourages specialization and ensures that overall investment levels are linked to the prevailing forecast of demand. Expansion programs are in effect allocated among the firms within the industry. Such behavior would violate the antitrust regulations enforced in the United States.

The Adoption of Major Technological Innovations

The extent to which investment by U.S. steel firms failed to maintain the industry's technological advantage is illustrated by the rate at which the two major steelmaking innovations of the postwar period, the BOF and continuous casting, were implemented.

The Basic Oxygen Furnace. The basic oxygen process is a refinement of the traditional Bessemer (or pneumatic) method of steelmaking. In the BOF, oxygen reacts violently with the charge in the furnace, drastically accelerating the refinement process and generating sufficient heat energy to eliminate the need for external fuel sources. Heat times

Table 3–2. Adoption of New Technologies, Various Countries (percent share and millions of net tons).

	U.S.		*Japan*		*EEC(9)*		*Canada*	
	%	*tons*	%	*tons*	%	*tons*	%	*tons*
A. *BOF*								
1960	3.4	3.3	11.9	2.9	1.6	1.8	28.1	1.6
1965	17.4	22.9	55.0	24.9	19.4	24.3	32.3	3.3
1970	48.1	63.3	79.1	81.2	42.9	65.1	31.1	3.8
1975	61.6	71.8	82.5	92.9	63.3	87.2	56.1	8.0
1981	60.6	73.2	75.2	84.1	75.1	103.6	58.6	9.4
B. *BOF plus Electric Furnace*								
1960	11.8	11.7	32.0	7.1	11.5	12.4	40.4	2.3
1965	27.9	36.7	75.3	34.1	31.5	39.5	45.1	4.6
1970	63.5	83.5	95.9	98.4	57.7	87.6	45.9	5.6
1975	81.0	94.5	98.9	111.3	82.6	113.7	76.4	10.9
1981	88.8	107.3	100.0	111.9	98.6	136.0	86.5	13.9
C. *Continuous Casting*								
1971	4.8	5.8	11.2	11.0	4.8	6.7	11.5	1.4
1976	10.5	13.5	35.0	41.4	20.1	29.7	12.0	1.7
1981	21.1	25.3	70.7	79.0	45.1	62.3	32.2	5.3

Source: International Iron and Steel Institute (IISI), *Steel Statistical Yearbook* (Brussels, IISI: various years).

in the early BOFs were roughly one-tenth those required in the average open hearths of that period (one hour vs. twelve hours).

The first BOF was a small prototype built by the Austrian firm Voest-Alpine in 1952. It was introduced into the United States in 1955 by McLouth Steel, a small integrated producer which filed for bankruptcy in 1981. Larger U.S. steelmakers remained skeptical, partially because they had not developed the technology, partially because they were unaccustomed to using pneumatic techniques, and partially because they were loathe to write off 40 million tons of new open-hearth capacity.

Nonetheless, the U.S. industry began its transformation to the BOF by the late 1950s. Although this was fairly late by world standards, American steelmakers adopted the new process very rapidly once they had committed themselves. This is shown in Table 3–2, which describes the adoption of the BOF in several countries or regions, both in terms of total tonnage and in terms of the share of total steel production. The data on BOF share can be misleading, since the residual can represent an efficient technology (if it is produced in electric furnaces) or inefficient technologies (if it is produced in open hearths or traditional pneumatic converters). For that reason, Table 3–2 also includes information on the tonnage and share of BOFs and electric furnaces combined; this then describes the extent to which a steel industry uses efficient steelmaking technologies.

Table 3–2 shows that the U.S. industry has lagged the Japanese in the adoption of the BOF and in the elimination of inefficient steelmaking capacity. Until recently, however, the United States had outperformed the Europeans, and U.S. steelmakers still use a higher proportion of efficient techniques than their Canadian counterparts. Furthermore, through the 1960s the smaller U.S. BOF and electric furnace share relative to the Japanese was linked to the much greater size of the U.S. industry. In terms of absolute tonnages, the United States installed BOFs at a rate comparable to that of the Japanese until the late 1960s.

Since 1975, the rate at which the American industry has eliminated open hearth capacity has been disappointing. Several U.S. mills—e.g., U.S. Steel's Fairless Works in eastern Pennsylvania and Geneva Works in Utah—are still open hearth plants. Improvements in open hearth technology (e.g., the use of oxygen) and the open hearth's ability to take more scrap than the BOF have enabled such facilities to remain barely viable, but they are clearly at a severe competitive disadvantage. Whereas Japanese and European steel producers have effectively eliminated the outmoded steelmaking technology, it retains a significant position within the North American industry.

Continuous Casting. Steel production has traditionally integrated a complex sequence of discrete steps, from the preparation of raw materials to the rolling of the finished product. As a result, steelmaking involves a high degree of waste due to the logistics of combining discontinuous steps into an overall flow: the processing of intermediate products involves significant yield losses; energy efficiency is reduced as intermediate products are cooled and then reheated for further processing; and labor productivity is reduced due to the handling of semifinished shapes. A more continuous process flow makes it possible to reduce such inefficiencies, and that is the appeal of continuous casting—a major step toward the fully continuous production of steel products.

Continuous casting replaces at least two of the process steps characteristic of traditional steelmaking methods: ingot casting and primary rolling. Molten steel is traditionally poured into ingot molds, where it is then cooled into heavy rectangular ingots. These can then be stored until needed, at which point they must be reheated and rolled into semifinished shapes: slabs, blooms, or billets. This stage is referred to as primary rolling. These semifinished shapes are stored until needed, reheated, and rolled into final steel products (slabs into sheets and plates, billets into bars and wire rods, and blooms into billets or various other products, e.g., tubes). With continuous casting, on the other hand, molten steel is poured directly into a mold through which it flows as it cools, until it leaves as a semifinished shape suitable for rolling into finished products.

This technique has several advantages. First and foremost, it greatly increases yield. Before ingots are rolled into finished shapes, their ends must be lopped off, producing scrap waste. This loss is effectively eliminated by continuous casting. Second, continuous casting reduces energy requirements, since it limits the number of times intermediate products must be reheated for further rolling. Third, labor productivity is increased, since several process steps, as well as the associated material handling, are eliminated. Finally, continuously cast steel is of better quality than steel made from ingots; metallurgy is more uniform and surface defects are reduced.

Continuous casting began to be adopted in the mid-1960s. Its first applications were to the production of simple shapes (bars and rods) from billets, since billets have a basically square cross section and are thus more easily cast than slabs. By the early 1970s, however, the technology for slab casting had been developed (although it still can be a finicky process); plants that did not employ this technology were at an increasingly serious disadvantage.

Table 3–2 shows the rate of adoption—in tonnages and in shares—of continuous casting for various countries and regions. In contrast to the record described for the BOF, the transition to continuous casting has been agonizingly slow in the United States. The extent to which the Japanese lead the United States in the application of this technology is substantial, and the European advantage is small only by comparison. While the United States kept pace with the rest of the world in the adoption of the 1960s technology (the BOF), it has been unable to maintain its position in the major new technology of the 1970s. Lack of funds, mandated expenditures for environmental purposes, deteriorating technical abilities, and other factors have combined to reverse the U.S. industry's traditional technological leadership.

Changes in Scale

The deterioration in the technological status of the U.S. steel industry has been matched by the elimination of the industry's scale advantage.

Table 3–3. Average Capacity of large[a] Plants and Firms in 1982.

	U.S.	*U.K.*	*Japan*	*France*	*Canada*	*Germany*
A. *Plants*						
Average Capacity[b]	3.8	4.2	10.4	4.4	4.0	4.8
Percentage of U.S.	100	109	273	116	105	128
B. *Firms*						
Average Capacity[b]	16.9	26.7[c]	33.2	4.6	5.0	9.8
Percentage of U.S.	100	158	196	27	29	58

a. Counting only the largest plants and firms up to two-thirds of total national capacity.
b. Millions of net tons.
c. British Steel Corporation (nationalized).
Sources: J.F. King, *World Capacity Report Quarterly Update* (Newcastle upon Tyne, July, 1982) and authors' estimates.

In the 1950s, the large scale of U.S. plants and firms was the most salient feature of America's dominant position in the world steel industry. By 1981, this advantage had disappeared, as is shown in Table 3–3. The scale of Japanese plants now sets the world standard; their average capacity is almost three times that of large U.S. integrated mills. Even European plants are significantly larger than those in the United States. Although only half the size of Japanese firms, American firms are still larger than the international average, but this is not directly relevant to physical efficiency. Only 21.5 percent of U.S. capacity is in plants whose scale exceeds five million metric tons. This is less than for any other country listed in Table 3–3; in Japan, the corresponding figure is almost 65 percent.[15]

Not surprisingly, the U.S. industry has lost its scale advantage at the facility level as well as at the plant level. This is shown in Table 3–4, which compares facility capacities in Japan and the United States for selected years. With the exception of the electric furnace, where the impact of nonintegrated producers is evident, every facility listed in this table shows the same pattern: a significant U.S. advantage in 1958 had been eliminated by 1980. In primary processes (ironmaking and steelmaking) and in the hot strip mill, this occurred much earlier—by 1964 in the case of blast furnaces.

Table 3–4. Average Capacity (output) of Facilities: Japan vs. U.S. (thousands of tons per year).

	1958		*1964*		*1972*		*1980*	
	Japan	*U.S.*	*Japan*	*U.S.*	*Japan*	*U.S.*	*Japan*	*U.S.*
Blast furnaces[a]	300.4	384.2	553.6	527.0	1494.4	653.2	2305.4	840.0
BOF	198.4	353.0	506.8	683.0	1278.8	1000.0	1647.9	1160.0
Electric furnaces	14.1	53.0	41.8	69.0	93.1	115.0	165.0	200.0
Hot Strip mills	661.4	1370.0	1594.0	1760.0	2491.2	2080.0	3344.4	2150.0
Plate mills	126.8	363.0	191.8	497.0	537.2	564.0	1277.6	646.0
Wire rod mills	111.3	221.4	175.0	257.0	274.5	312.0	490.5	415.1
Cold Reduction mills	170.9	523.0	208.3	584.0	357.1	685.0	784.8	699.0

a. Output rather than capacity.
Sources: American Iron and Steel Institute (AISI), *Directory of Iron and Steel Works in the U.S. and Canada* (New York and Washington, D.C.: AISI, various years); Japan Iron and Steel Federation (JISF), *Statistical Yearbook* (Tokyo: JISF, various years); and Japan Ministry of Labor (JML), *Labor Productivity Statistics Survey* (Tokyo: JML, various years).

Research and Development

As the data presented above indicate, the American steel industry—at least its integrated sector—has now lost the technological advantage which provided the foundation for the industry's competitiveness in the first two decades after World War II. To some extent, this can be linked to the more favorable economic conditions, especially in terms of market growth, enjoyed by foreign producers. Yet U.S. firms failed to implement strategies that could have compensated for the shifting pattern of world production by building on the strengths of the American industry. This is evident in the industry's weak commitment to technical research and development. When compared with other U.S. industries, steel devotes a paltry share of revenues to scientific research. Steel R&D expenditures have been declining as a share of industry revenues over the past twenty years. From 1975 to 1980, less than 0.6 percent of the industry's net sales revenue was devoted to R&D; this places steel among the lowest of the major industry groups for which such data are kept. By comparison, almost 10 percent of net sales were devoted to research and development in the computer industry over this period, 4.4 percent in the machinery industry, and 0.9 percent in the nonferrous metals industry. Steel's relative indifference to technological performance is also evident in the personnel employed by the industry. According to the National Science

Foundation, ferrous metals has a smaller proportion of R&D scientists and engineers in its work force (0.5% in 1977) than any other major industry group except textiles.[16]

Furthermore, most R&D in the steel industry is oriented toward product development, for which the industry has an impressive record of technological achievement. Yet this emphasis has shortchanged basic and applied research, which have the potential of developing new technologies for improving efficiency and reducing costs. Less than 20 percent of the industry's R&D effort is devoted to basic and applied research.

This lackluster commitment to steel research and development has disturbing implications: new process technologies are not being adequately developed in the United States. Both the BOF and continuous casting were invented and refined in Europe, in spite of the supposedly dominant status of the American industry at the time they were introduced. This pattern suggests that it will be extremely difficult for American steel producers to reverse the deterioration in their technological competitiveness. Due to their own failure to invest in process research, American firms are increasingly turning to foreign producers for operating technologies. In one sense this represents an improvement over the technological ethnocentricity that formerly characterized the industry. In a more significant sense, however, it threatens to leave the industry perpetually behind. By the time U.S. producers purchase a new technology, its developer has already achieved efficiency levels and technical refinements that are unavailable to the purchaser. The indifferent commitment to R&D on the part of the U.S. steel industry, while often hidden from view, is one of its principal weaknesses, a legacy from the days when the industry was a price maker and hence did not have to concentrate on aggressive cost competition. Unfortunately for U.S. integrated producers, this is a legacy that is not shared by their competitors. Japanese steel firms allegedly devote 1.6 percent of net sales revenue to R&D.[17] Such discrepancies have generated predictable results.

THE LOW-COST PRODUCER NO LONGER

Firms and industries compete on the basis of costs. Thus, all of the themes that have been discussed to this point must finally be woven into the fabric of comparative costs. Technological trends, changes in demand, levels of investment—all of these issues are irrelevant if they do not affect costs. While profits are the measure of success in a private economy, profitability is presumably determined by comparative costs. We will conclude this chapter with a brief summary of the deteriorating

Table 3–5. Comparative Integrated Production Costs in 1981: Cold-rolled sheet[a] (dollars per net ton shipped, normal operating rates[b]).

	U.S.	*Japan*
Labor	144	71
Iron ore	63	50
Purchased Scrap	16	–
Coal or coke	54	61
Other energy	54	44
Other[c]	83	82
Total operating costs	414	308
Depreciation	18	27
Interest	7	30
Miscellaneous taxes	7	5
TOTAL COSTS[d]	445	370

a. Cold-rolled sheet, in coiled form, at 1981 exchange rates.
b. Operating rates are an average of those between 1977 and 1981. U.S. operating rates averaged 80 percent over that period; Japanese, 65 percent.
c. Includes alloying agents, fluxes, refractories, contracted services, and so on.
d. Excluding any return on equity.
Sources: Estimated by authors from data contained in World Steel Dynamics, *Core Report Q* (New York: Paine Webber Mitchell Hutchins, 1982); company annual reports; Department of Commerce, *TPM Manual* (Washington, D.C.: U.S. Department of Commerce, various issues). See Appendix. C

profitability of American steelmakers, but this will be preceded by a lengthier discussion of cost trends.

Table 3–5 presents estimates of total 1981 costs for the production of cold-rolled sheet (a representative integrated product) in the United States and Japan. These estimates concur with the conclusions reached by other studies that have addressed this issue.[18] Japanese integrated steel firms enjoy a significant advantage vis-a-vis their U.S. rivals in terms of operating costs, and this advantage is not offset by higher Japanese capital costs. The costs of entering the U.S. market may boost Japanese costs by roughly 17.5 percent (about $65 a ton in 1981), but this final barrier no longer maintains the cost advantage of the U.S. industry. The U.S. industry has generally been cost competitive with European producers, sometimes enjoying a significant advantage. But even here the trends suggest a reversal of this relationship by the mid-1980s. If the exchange rates prevalent in mid-1982 are sustained, the U.S. industry has already fallen behind all its major European competitors in

Figure 3–6. Comparative Integrated Total Production Cost Trends; Cold-rolled Sheet ($/net ton).

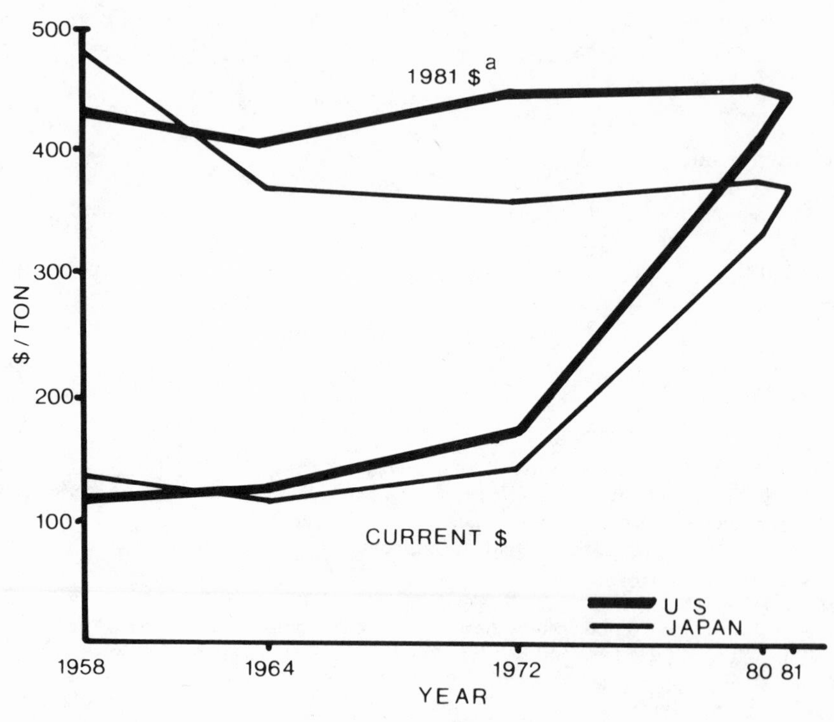

a. Converted from current to 1981 dollars using the Producer Price Index for Industrial Products of the U.S. Department of Labor, Bureau of Labor Statistics.
Sources: As in Table 3–5.

terms of dollar denominated costs. Exchange rate adjustments may reverse this present relationship, but they will not reverse the underlying trends. Costs are increasing more slowly abroad than in the United States.

The historical pattern of U.S. and Japanese cost competitiveness is described in Figure 3–6, which is based on estimates of total costs for cold-rolled sheet and which describes cost trends in both current and constant (1981) dollars. Figure 3–6 shows that the U.S. industry has consistently failed to reduce its real costs, in spite of gradual increases in productivity and efficiency. The Japanese industry, on the other hand, was able to reduce real costs significantly in the late 1950s and early

1960s. Even in terms of nominal costs, given inflationary pressures, Japanese costs in 1972 were no higher than in 1958; in the United States, on the other hand, nominal costs had grown by over 40 percent during the same period. Since 1972, however, real costs seem to have risen faster in Japan than in the United States. While this is chiefly due to exchange rate fluctuations, it also stems from the fact that the Japanese industry, having already reached a high level of efficiency, has not been able to sustain the progress it made prior to the mid-1970s. More ominously, this suggests that the Japanese industry has been unable to cope with the global steel crisis. It is possible that the Japanese are now experiencing the kind of dilemma which characterized the decline of the American steel industry from a position of comparable preeminence.

The curves in the lower part of Figure 3–6 show the spectacular cost increases incurred by both the U.S. and Japanese steel industries after the energy crisis in 1973. This reflects the general inflationary environment, but steel costs have risen more rapidly than the overall rate of inflation, as have steel prices. This is not surprising given the energy-intensive nature of steel production; the implications of this trend, however, are more profound than would appear at first glance.

The fact that prices and costs in the steel industry are rising more rapidly than the rate of inflation encourages substitution away from steel, thus favoring further inroads by alternative materials. It is true that some of the materials that compete with steel (e.g., aluminum and plastics) are also energy intensive, but these products are much lighter; their properties are therefore more suited to a world of higher real energy costs. The energy crisis thus spurred a further deterioration in steel intensity. Second, rapid increases in real energy costs make it essential for energy-intensive industries to improve energy efficiency—a goal that entails significant capital expenditures. Since the American steel industry was already operating under severe capital constraints, its ability to reduce energy costs was limited relative to industries like Japan's. This is reflected most starkly in the differing rates at which these two industries have adopted continuous casting technology (see Table 3–2). Thus it seems likely that the cost pressures generated by increasing real energy costs and a general inflationary environment portend further competitive problems for American steel producers on two fronts: relative to competing materials and relative to competing steel suppliers.

In spite of comparable real increases in dollar-denominated costs since the mid-1970s, Japanese steelmakers still enjoy a significant cost advantage vis-a-vis U.S. producers. If this result is evaluated from the perspective of the American industry, two factors stand out as having directly "caused" the decline of U.S. cost competitiveness. First, the technological advantage of the U.S. industry has been eliminated. Sec-

Table 3–6. Estimated Operating Costs: Cold-rolled Sheet, Japan versus U.S.

	Price of Input ($)[a]		*Use/Net Ton CRS*		*Unit Cost ($/ton)*	
	U.S.	*Japan*	*U.S.*	*Japan*	*U.S.*	*Japan*
1964						
Labor	4.35	0.90	10.01	19.12	44	17
Iron ore	12.00	12.85	1.44	1.72	17	22
Purchased scrap	34.60	38.20	0.20	0.08	7	3
Coking coal	9.25	14.65	0.85	0.94	8	14
Other energy					15	16
Other[b]					26	30
Total materials (+ other)					73	85
Total operating costs					117	102
1972						
Labor	7.10	2.75	8.07	8.21	57	23
Iron ore	14.55	10.90	1.58	1.86	23	20
Purchased scrap	30.83	36.75	0.15	–	5	–
Coking coal	15.74	20.00	0.93	0.88	15	18
Other energy					20	17
Other[b]					35	35
Total material (+ other)					98	90
Total operating costs					155	113
1980						
Labor	18.80	11.00	7.20	5.84	135	64
Iron ore	36.00	25.50	1.59	1.81	58	46
Purchased scrap	89.50	100.00	0.16	–	14	–
Coking coal	52.50	65.00	0.85	0.89	45	58
Other energy					46	42
Other[b]					76	76
Total materials (+ other)					239	222
Total operating costs					374	286

a. Dollars per hour for labor, dollars per ton for iron ore, scrap, and coking coal.
b. Refractories, rolls, some contracted services, and so on.
Sources: World Steel Dynamics, *Core Report J* and *Core Report Q* (New York: Paine Webber Mitchell Hutchins (1976 and 1982); *Tex Reports* Tokyo: various years); Federal Trade Commission (FTC), *Staff Report on the U.S. Steel Industry and Its International Rivals* (Washington, D.C.: U.S. Government Printing Office, 1977); and author's estimates.

ond, the United States has lost ground relatively in terms of raw materials costs. Stagnating productivity has transformed high employment costs into a serious disadvantage, and other countries have reduced material costs at a faster rate. We now turn to these themes.

Trends in Material Costs

Table 2–4 presented estimates of 1958 basic operating costs in the U.S. and Japanese steel industries, and this analysis is extended to later years in Table 3–6. The data presented there show that Japanese firms have been able to reduce the impact of relatively high materials costs, especially for iron ore. In the late 1950s, Japan's status as a resource poor country still limited its competitiveness in the world steel market. By the late 1960s, however, this disadvantage had been reversed. Increased efficiency had lowered materials costs per unit of output. More surprisingly, however, the Japanese had also been able to reduce the price they paid for imported raw materials. By the early 1970s, the relatively rich resource base of the United States was no longer an unambiguous advantage for American steel producers.

The Japanese effort to alleviate the burden associated with high materials prices was based on reducing transportation costs and on securing long-term supplies at the lowest possible price. Nowhere was this effort more successful than with iron ore. Since the Japanese industry did not have the option of relying on domestic ores, it was able to view its iron-sourcing problem strictly as a cost question, pioneering the use of low-cost ores from new sources, shipped in huge bulk carriers. The success of the Japanese raw materials strategy transformed a 28 percent disadvantage in 1958 (see Table 2–4) into a 30 percent advantage in 1980.

The U.S. steel industry has relied on North American sources of iron ore; production costs for these ores are significantly above those in other major producing countries—e.g., Brazil and Australia—for equivalent grades and types of ores (see Appendix B). In general, U.S. producers have valued ownership of iron mines more than they have valued cost advantages in the acquisition of ore. The vertically integrated structure of major U.S. firms has made it difficult for them to allocate total costs correctly among different operations and therefore to alter practices according to market signals, such as the lower prices of foreign ore. Overly optimistic assumptions about potential iron ore shortages, tax considerations (e.g., the desire to gain depletion allowances), inertia, and managerial empire building have encouraged U.S. firms to maintain their commitment to domestic sources of iron ore.

Despite major reductions in shipping costs since the early 1960s, transport costs continue to be a major obstacle to the use of Brazilian or Australian ores in the United States. North American ores have been cheaper—or only slightly more expensive—at most U.S. steel mills, particularly in the Great Lakes. But this has less to do with any inherent cost advantage than with the fact that U.S. mills are poorly located by world standards. The inland sites occupied by most major U.S. mills offer the advantage of proximity to markets but are a major handicap in terms of iron ore costs. Because of the vertically integrated structure of major U.S. steel firms, even plants on the Atlantic or Gulf coasts rely too heavily on high-cost North American ores.

The relative progress made by the Japanese in reducing iron ore costs is the most dramatic example of a pattern that applies to most raw materials. For scrap and coking coal, the United States has retained a price advantage due to abundant domestic reserves of these inputs. Yet, in each case the relative Japanese disadvantage has narrowed considerably since the late 1950s: from 84 percent to 24 percent in coking coal and from 27 percent to 12 percent in scrap. Here as well, significant real reductions in shipping costs and expansion in transoceanic trade has benefited industries that rely on international markets, such as the Japanese, relative to industries that have tended to pursue a more autarchic status, such as the United States. When combined with Japan's greater efficiency (suggested by the usage column in Table 3–6), comparative trends in the price paid for inputs had led to a reversal in the traditional U.S. advantage in basic materials costs by the early 1970s. Although Japanese materials costs had been 40 percent above U.S. levels in 1958 (see Table 2–4), by 1980 they were roughly 6 percent below those incurred by the American industry.[19]

Employment Costs

High labor costs are now invariably cited as one of the most difficult problems, if not *the* most difficult, faced by American steel producers. This seems to be confirmed by Table 3–7, which presents historical data on dollar denominated employment costs in several steel industries; it shows that the U.S. industry faces the highest employment costs in the world. As one would expect, the differential vis-a-vis industries in developing countries is even greater; employment costs in the Korean steel industry, for instance, are estimated to be less than one-tenth those prevalent in the United States. Since the costs described in Table 3–7 are dollar denominated, they reflect shifts in exchange rates as well as

Table 3–7. Hourly Employment Costs, Various Countries (in current dollars).

	U.S.	*Japan*	*W. Germany*	*France*	*U.K.*
A. Absolute Levels (in current dollars)					
1969	5.54	1.65	2.36	2.19	1.66
1972	7.33	2.86	4.24	3.46	2.62
1975	10.83	5.54	7.61	7.23	4.57
1978	14.73	9.44	11.55	10.56	5.93
1981	20.78	11.57	13.18	12.65	9.56
1982 (p)	24.42	11.03	13.35	12.39	9.23
B. Annual Rates of Growth, 1969–81 (percent)					
In dollars	11.6	17.6	15.4	15.7	15.7
In home currency	11.6	13.0	10.2	16.2	17.3

Source: World Steel Dynamics, *Core Report Q* (New York: Paine Webber Mitchell Hutchins, 1982).

actual trends in employee compensation. For that reason, data on the growth rate in home currency employment costs are also presented.

Although the most striking feature of the data presented in Table 3–7 is the extent to which hourly employment costs in the United States have persistently exceeded those faced by other steel industries, these statistics also show that the *relative* U.S. disadvantage has declined. Whereas U.S. employment costs were eight times the Japanese level in 1956, they are now roughly twice as high.[20] This relative convergence, however, was associated with an absolute increase in the U.S. disadvantage. Continuing the comparison with Japan, the U.S. premium in employment costs increased from $2.92 in 1956 to $9.18 in 1981. This increased differential actually represents a slight decline in real dollars. Judged in real terms, the disadvantage faced by the U.S. industry in employment costs does not appear to have increased either absolutely or relatively.

Nevertheless, relative constancy in the real labor-cost disadvantage faced by American steel producers actually represents a significant deterioration in the competitiveness of the U.S. industry. Given the fact that Japanese productivity is now significantly greater than the level achieved by the U.S. industry, unit labor costs are an increasingly severe disadvantage.

Figure 3–7. Hourly Employment Costs in the Steel Industry as a Percent of the Manufacturing Average.

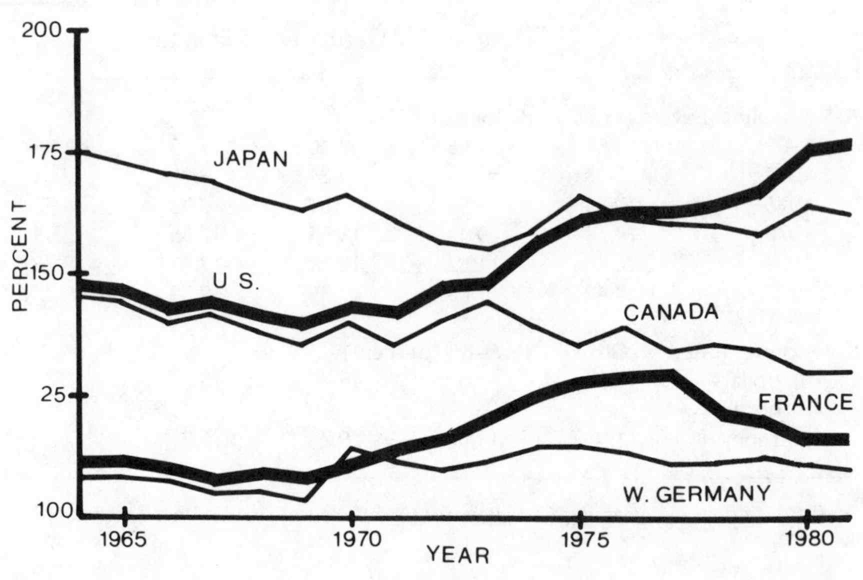

Source: U.S. Department of Labor, Bureau of Labor Statistics.

The extent to which increased employment costs have undermined the international competitiveness of the American steel industry can also be discussed in terms of the relationship between employment costs in steel and in other American industries. Since the establishment of floating exchange rates in 1973, currency fluctuations should adjust to varying national trends in productivity and unit labor costs. Relative international wage rates should follow trends in productivity, allowing exchange rate shifts to support continued trade while boosting living standards in those countries where productivity growth has been strong. Accordingly, over the past two decades dollar–denominated wage rates have grown more slowly in the United States than in countries where productivity growth has been more robust—e.g., Japan. A relative leveling of average employee compensation levels in advanced countries has been particularly evident since the late 1960s.

Figure 3–7 shows the differential between steel employment costs and the manufacturing average in several countries from 1964 to 1981. Steelworkers are generally paid a premium, partially because of the difficult nature of their work and partially because the steel sector tends to be characterized by a small number of large firms and by strong unions. This combination traditionally encourages relatively high wages. Yet this graph shows that the relative standing of the U.S. steelworker has increased rapidly in the 1970s and 1980s. During the 1960s, the premium paid by the U.S. steel industry was roughly comparable to world standards; in fact, the relative standing of U.S. steelworkers actually decreased during that decade. In 1972, however, the U.S. steel premium began a sharp ascent. By the late 1970s, the differential between steel compensation and the manufacturing average was greater in the United States than in any other country; and this trend has accelerated since then. While the data presented in Figure 3–7 indicate that the Japanese premium is close to the U.S. level, this may well be incorrect. Forty to 50 percent of the labor force in Japanese steel mills is composed of contract workers, who are paid less than regular employees. It is probable that the data used to construct Figure 3–7 underestimate the proportion of contract workers in Japan and thus overstate employment costs for Japanese steel producers.

Clearly, the premium paid to steelworkers in the United States is a severe competitive disadvantage for the domestic industry. Whereas employment costs in other American manufacturing industries tended to fall relative to foreign levels as the postwar productivity advantage of the U.S. economy eroded, this was not the case in steel. Unlike their counterparts in other industries, steelworkers were able to increase their standard of living in a decade of general economic retrenchment. It seems particularly perverse that this differential began to increase most dramatically just at the time the industry lost the productivity advantage that could support high wage rates. The principal reason for this was the Experimental Negotiating Agreement (ENA), a novel labor contract concluded in 1973 and originally viewed as a means of combating imports.

During the early 1960s, steel imports surged periodically during years in which contracts were negotiated between the industry and the United Steelworkers. This can be seen in Figure 3–4, where contract years are listed below the chart; these roughly coincide with major boosts in import penetration. This correlation suggested that imports benefited from the inventory expansion undertaken by consuming industries as a hedge against the possibility of a steel strike. Besides its impact on trade flows, hedge buying by steel consumers fearing a steel strike also contributed greatly to cyclical volatility in steel demand. If a strike did not occur (and none occurred industry wide after 1959), excessive inventories had to be drawn down, so that demand fluctuated sharply around each contract negotiation.

For these reasons, the industry placed a high priority on eliminating the potential for steel strikes; it was felt that this would turn the tide against imports and smooth out the order book. As a result, the industry and the union developed the ENA for the 1974 contract. This agreement guaranteed steelworkers substantial wage increases in exchange for their relinquishing the right to strike. Unfortunately, by the time the ENA was concluded, the advantages enjoyed by many foreign steel firms were based squarely on lower costs rather than on the factors that were addressed by the agreement. In the inflationary environment of the 1970s the agreement exacerbated what was becoming the chief liability of the U.S. industry: high costs of production.

In effect, the industry believed its own propaganda; namely, that the periodic threat of a steel strike provided the foundation for ratchetlike surges in the import share. Under the ENA, nominal compensation for steelworkers increased 260 percent from 1973 to 1981; the consumer price index increased 204 percent during the same period. Most of the steel increase was in nonwage benefits (which increased by 344%) and in a cost-of-living allowance.[21] In principle, the industry cannot be faulted for the failure to anticipate the endemic inflation that developed after the ENA was implemented. Yet it renewed the agreement twice. After a long period of silence,[22] it has now begun to address its disadvantage in employment costs by blaming the union, which was only one of the two signatories to the ENA.

In the extremely depressed market conditions of 1982, the contract coordinating committee representing the eight major integrated firms sought to renegotiate its existing contract with the United Steelworkers. Although this effort failed twice–the second time despite support by the union leadership for the proposed contract revisions–a new contract was negotiated in March 1983. While the eventual financial effects of this contract cannot be calculated in advance, its principal feature is an approximate 9 percent reduction in employment compensation. This is to be gradually restored over the life of the contract. In exchange for this concession, the companies agreed to boost supplemental unemployment benefits and to reinvest the savings provided by the contract in existing steel plants.

Givebacks of this magnitude do not appreciably affect the competitive disadvantage associated with existing employment costs (see Table 3–7). Insofar as competitive pressures force steel wages and benefits down, this is increasingly likely to occur outside the traditional pattern of collective bargaining–i.e., industrywide negotiations between the union and the coordinating committee, dominated by U.S. Steel. One example of this is provided by local negotiations over work rules (e.g., reducing work crews or combining maintenance functions). Altered work rules may boost productivity significantly and are therefore likely to have a greater impact on unit labor costs than marginal reductions in hourly compensation.

A more important trend undermining the industrywide labor contract is the prevalence of agreements at individual plants or companies that provide more substantial reductions in compensation than those included in the general contract. Such agreements were concluded in 1982 at McLouth and Wheeling-Pittsburgh. At McLouth, major wage concessions and streamlined work rules were essential components in the purchase of the then bankrupt firm by Cyrus Tang. Wheeling-Pitt, also a relatively vulnerable firm, obtained wage concessions in exchange for job security commitments and a profit-sharing plan. The most intriguing instance of this trend was provided by the employee purchase of National Steel's Weirton Plant, where workers agreed to a 30 percent reduction in wages as part of the purchase agreement. These agreements are certain to increase the competitive pressure on firms which are covered by the more expensive industrywide contract. This in turn is likely to lead to a cascading pattern of local concessions in exchange for some measure of job security, undermining industrywide pattern bargaining.

The employee purchase of Weirton Steel may be a harbinger of future exit from the industry; the tax advantages of selling a plant to its labor force make such an approach an attractive means of divestiture for companies seeking to reduce their involvement in steel. It is less clear that employee purchase can actually save any jobs in the long run. Instead, it may eventually encourage subsidies by states (West Virginia, in Weirton's case) anxious to retain employment—a development which, like foreign subsidies, would further distort the operation of the market mechanism in the steel industry. At any rate, it is clear that the existing pattern of industrial relations in the steel industry is now fragmenting, as is the industry itself.

Profitability

Weak markets, foreign competition, lagging investment, and reduced cost competitiveness have all contributed to increasing pressure on the profit margins of American steel firms. In the final analysis, profits reflect and define the competitive prospects of any industry. While providing a performance standard in any given period, profit rates also determine the future performance of an industry by establishing whether capital will flow into or out of it.

Table 3–8 compares steel industry profit rates with the manufacturing average for various periods since 1950. It shows that the profit rate

Table 3–8. Return on Equity, Steel vs. All Manufacturing, 1950–1981. (%).

	Steel Industry	*Manufacturing Average*
1950–54	11.6	13.7
1955–59	11.9	12.7
1960–64	7.2	11.1
1965–69	8.1	13.3
1970–74	8.1	12.6
1975–81	7.7	15.6

Sources: American Iron and Steel Institute (AISI), *Annual Statistical Report* (New York and Washington, D.C.: AISI, various years); and Citibank.

in steel has failed to reach the manufacturing average in every period. As was mentioned earlier, steel profits have exceeded the manufacturing average only once since the 1950s, in 1974. Despite the recentness of this event, the relative performance of the steel industry has gradually deteriorated since the 1950s, when steel's return on equity approached the manufacturing average. From 1975 to 1981, the rate of return in steel was less than half the average earned by U.S. manufacturing industries. Furthermore, even this overstates the profitability of steel operations, since the data in Table 3–8 include the results of the nonsteel operations that are becoming increasingly important to the firms that report profitability to the American Iron and Steel Institute. While the equity of steel firms cannot be disaggregated into steel and nonsteel components, the greater profitability of nonsteel operations is apparent in the annual reports of most of the firms that have diversified away from steel.[23] Finally, Table 3–8 does not include financial results for 1982, when steel firms, including nonsteel operations, lost about $3.3 billion—17 percent of stockholders' equity at the end of 1981.

The low rate of return in steel is the principal determinant of the industry's inadequate investment; at the same time, lagging investment ensures that poor profits will persist. The long-standing failure of steel profits to reach the manufacturing average suggests that further contraction of the industry is assured. For steel executives, low profitability is largely the fault of government policy, especially the political failure to restrain imports—although tax policies, regulatory policies, and so on have also been cited by steel firms to explain their poor performance. Regardless of the policy environment, however, managerial performance must be evaluated according to the standard of profitability. On that basis, steel firms have clearly failed to cope with the competitive environment of the postwar period, and this is the reason they have had difficulty raising capital for investment.

CONCLUSION

This chapter has provided a brief description of the American steel industry's fall from the position of unrivaled world supremacy that was documented in Chapter 2. Evidence of that decline has been apparent since at least the early 1960s, when the industry still enjoyed tremendous advantages in performance, scale, and resources. By the 1970s, this was no longer the case. The industry was caught in a vicious cycle of weak demand, poor profitability, lagging investment, and high costs. Its conditions had changed drastically, yet no strategy emerged to cope with these changes.

The most fundamental change confronting traditional U.S. steel firms has been an absolutely fundamental erosion in the barriers to entry in the steel market. For most of the twentieth century, the steel market in the United States had been supplied by a relatively stable oligopoly. This entire fabric began to unravel in the 1960s. The increased availability of competitive imports was the most visible form of entry, but there were significant changes in the structure of the domestic industry as well—a topic that will be discussed in the next chapter.

Competition from new entrants has exacerbated the problem of declining steel intensity and the resultant weakness in demand. This has encouraged the U.S. industry to concentrate its attention on reducing competition by seeking trade restraints. Regardless of the moral status of the industry's complaints, its efforts to restrict imports have had paltry effects. Perversely, the failure to stem the import tide has elicited ever more frantic lobbying efforts on this issue, diverting attention from the risks inherent in staking the industry's future on its political clout. Instead of seeking to prop up a disappearing status quo, integrated firms should be implementing strategies that are based on the contemporary realities of the world steel market. Declining steel intensity in advanced economies, and thus slow growth in demand, is is an unpleasant but undeniable fact, as is the presence of significant foreign competition.

Instead of developing a strategy attuned to these realities, integrated producers in the United States have responded defensively to their vulnerability on the cost front. They have been slow to adopt the newest technology, despite the fact that employment costs should have made this the highest priority. They have devoted a miniscule share of revenues to research and development, so that major process innovations have been developed abroad. Finally, they have failed to grasp the absolute imperative of dynamic cost competitiveness, maintaining increasingly disadvantageous sources of raw materials and trading wage restraint for the sake of a protectionist cabal in Washington.

If there were any lingering doubts, the catastrophe of the 1982 steel industry recession has shown clearly that the industry's priorities have been misplaced. As its competitive position eroded, so did the viability of its traditional pattern of behavior. Yet, an alternative strategy has still not emerged. The more this is delayed, the more radical it must be.

Notes

1. Speech given at the General Meeting of the American Iron and Steel Institute, New York, May 28, 1982.
2. Can Manufacturers Institute (CMI), "Year-to-date Metal Cans Shipment Report" (Washington, D.C.: CMI, various issues).
3. American Iron and Steel Institute (AISI), *Annual Statistical Report, 1971* (New York: AISI, 1972), p. 26; AISI, *Annual Statistical Report, 1981* (Washington, D.C.: AISI, 1982), p. 32.
4. James Cook, "The Molting of America," *Fortune,* November 22, 1982, p. 165.
5. AISI, *Annual Statistical Report 1971* and *1981*, pp. 26 and 32, respectively.
6. Office of Technology Assessment, *Technology and Steel Industry Competitiveness* (Washington, D.C.: U.S. Government Printing Office, 1980), p. 170.
7. International Iron and Steel Institute (IISI), *Steel Statistical Yearbook, 1981* (Brussels: IISI, 1981), p. 31.
8. Economic Commission for Europe, United Nations, *Long-term Trends and Problems of the European Steel Industry* (Geneva: U.N., 1959), p. 21.
9. Organization for Economic Cooperation and Development (OECD), *The Steel Market in 1981 and the Outlook for 1982* (Paris: OECD, 1982), pp. 42–74.
10. Ibid., pp. 42–74.
11. Economic Commission for Europe, *Long-term Trends and Problems,* p. 21; IISI, *Steel Statistical Yearbook, 1982* (Brussels: IISI, 1982), p. 3.
12. Economic Commission for Europe, United Nations, *The Steel Market in 1980* and *The Steel Market in 1970* (Brussels: U.N., 1981 and 1971, respectively), for capacity; IISI, *Steel Statistical Yearbook, 1981* and previous editions for consumption. Three-year averages

around 1970 and 1980 were used to calculate rates of growth in consumption.

13. See, for example, Kiyoshi Kawahito, *The Japanese Steel Industry; with an Analysis of the U.S. Steel Import Problem* (New York: Praeger, 1972).
14. AISI, *Annual Statistical Report* (New York and Washington, D.C.: AISI, various years).
15. J.F. King, *World Capacity Report Quarterly Update* (Newcastle upon Tyne: J.F. King, July 1982).
16. National Science Foundation, *Research and Development in Industry, 1980* (Washington, D.C.: U.S. Government Printing Office, 1982), pp. 24 and 39.
17. Kunio Okabe, "World Steel Industry Must Counter Stacked Deck," *American Metal Market*, November 9, 1982, p. 18.
18. See, for example, Council on Wage and Price Stability, *Prices and Costs in the U.S. Steel Industry* (Washington, D.C.: U.S. Government Printing Office, 1977) and Federal Trade Commission, *The United States Steel Industry and Its International Rivals* (Washington, D.C.: U.S. Government Printing Office, 1977).
19. The argument that the U.S. steel industry no longer enjoys an advantage in raw materials costs has been advanced more recently (and in greater detail than here) by Robert Crandall, *The U.S. Steel Industry in Recurrent Crisis* (Washington, D.C.: The Brookings Institution, 1981).
20. World Steel Dynamics, *Core Report J* and *Core Report Q* (New York: Paine Webber Mitchell Hutchins, 1979 and 1982, respectively); Bureau of Labor Statistics, U.S. Department of Labor.
21. AISI, *Annual Statistical Report, 1981* (Washington, D.C.: AISI, 1982), p. 22; U.S. Department of Commerce, "Survey of Current Business" (various issues).
22. Labor costs are not even mentioned in AISI's otherwise comprehensive report *Steel at the Crossroads* (Washington, D.C.: AISI, 1980).
23. One example of the impact of diversification on a major steel firm is described by Donald B. Thompson, "U.S. Steel's Chemical Stepchild Grows Up," *Industry Week*, June 23, 1980, p. 73.

4 SEEDS OF RENEWAL: STRUCTURAL CHANGE AND THE AMERICAN STEEL INDUSTRY

We are going to be predominantly a steel company well throughout the balance of this century.[1]

—David Roderick
Chairman, U.S. Steel Corporation
October 1978

The decline of the U.S. steel industry is but one product of the forces that are now reshaping the American economy and society, even though the precipitousness of the industry's decline can be partly attributed to ill-conceived industry strategies and government policies. Shifts in individual tastes and priorities are altering purchasing patterns. Basic industries are deteriorating, while high-tech sectors are growing at a rapid rate. Regional patterns of production, incomes, and population are being altered. Significant shifts in relative prices, although partially masked by high rates of inflation, have changed the competitive standards and prospects of traditional modes of production. Educational and skill requirements are changing with the industrial structure, disrupting traditional patterns of labor organization and relative wages.

The purpose of this book is not to delineate the general features of the industrial revolution that the American economy is now experiencing. Our concern is the implications of these changes for the steel industry—and, by analogy, for other basic industries as well. Two of these changes were introduced in the preceding chapter: shifts in demand leading to decreased steel intensity and the increasing involvement of the United States in a world economy. A third theme, the rapid tempo of innovation and changing techniques, provides the technological background to the trends that comprise the subject of this chapter.

For the economy as a whole, the development and widespread adoption of advanced electronics (data processing, robotics, etc.) are radically altering production, employment, and skill requirements. Small-scale light industry and work centers promise to replace large-scale heavy industries and massive office complexes, dispersing productive activities and fragmenting markets. The rapid adoption of new process technologies and new product orientations is becoming a competitive imperative for all industries.

Even in the traditional manufacturing sector, economic growth is likely to foster the dispersion of productive activity and to diminish the importance of scale economies. This tendency is already more advanced in steel than in other basic industries, as this chapter will show. The shift towards lighter, more flexible operations is being fostered by changes in the technology of steel production. Moreover, an environment of high real interest rates is accelerating this trend by tightening the capital constraints faced by the industry. The forces now reshaping the U.S. economy have placed inordinate demands on capital markets: growth industries must be financed, older industries must modernize and improve efficiency, extensive social programs must be funded, and so on. These requirements are raising the real opportunity cost of capital. Higher real interest rates will probably be with us for some time, as the economy copes with the necessity of extending individuals' time preferences (i.e., increasing the attractiveness of saving) and funding the required changes in the structure of economic activity. While capital shortages may be a new phenomenon to many industries, American steel producers have lived in a world of fairly stringent capital rationing since at least the 1960s. The way in which the steel industry attempted (or failed) to cope with this environment provides important lessons for other industries.

The structural changes occurring in the steel industry should be evaluated within the context of the more general forces that are now transforming American society. Developments in steel are reshaping the industry along lines that are more compatible with the emerging structure of the U.S. economy. The decline of the American steel industry is frequently discussed, yet there is still little recognition of the extent to which the gale of "creative destruction" (to use Schumpeter's phrase) is remolding the industry. The pressures which have led to overall degeneration have also called forth spectacular successes; the likely structure of a more competitive, more profitable industry has emerged as a result of the fragmentation induced by overall decline. If the past twenty years have for some steel firms been the worst of times, for others they have truly been the best.

STRUCTURAL CHANGES IN THE INTEGRATED SECTOR

Although they still constitute the predominant sector of the U.S. steel industry, integrated firms no longer enjoy the preeminence that they had maintained from the turn of the century to the 1960s. Integrated firms have been subject to increasingly intense pressures generated by changes in tastes, techniques, trade, and so on. They are losing their competitive position within the steel market, even as they see that market decline in importance for the U.S. economy. In 1955, the eight largest steel firms were all among the fifty-five largest U.S. corporations in terms of sales; twenty-five years later, only two could still make that claim based on steel sales.[2] Integrated firms have failed to respond aggressively to these pressures; their behavior has changed little in spite of the radically different competitive environment they face.

From the formation of the U.S. Steel Corporation in 1901 until the 1960s, the American steel industry represented a classic example of an oligopolistic industry, both in terms of structure and behavior. U.S. Steel Corporation, as the unchallenged industry leader, in effect determined the industry's overall price structure. Other large firms were generally more efficient, so that it was relatively easy for them to prosper under the umbrella of their unwieldy rival and leader. Price competition was as a rule banished from the world of Big Steel, and with it much of the incentive for reducing costs.

If the American steel industry once provided a textbook case of oligopoly, it has now vindicated the textbook principle that oligopoly attracts entry. In the early 1960s, foreign producers and mini-mills began to increase their presence in the American steel market. Accounting for less than 5 percent of U.S. consumption in the late 1950s, these producers met almost 40 percent of America's demand for steel in 1982.

This rapid restructuring of the U.S. steel market is illustrated in Table 4–1, which describes the post-1950 trend in market share for each of the eight largest firms as of 1981. These data clearly show the differentiated pattern of competitive performance in the U.S. steel industry. The erosion of market share is most pronounced among the largest firms, especially U.S. Steel. Such deterioration represents the continuation of a long-standing trend in the market share of the industry leader; U.S. Steel's share of total domestic production has fallen consistently, from over 54 percent in 1910 to less than 20 percent in 1981.[3] Unfortunately, it is impossible to establish which of several causes (diseconomies of scale, managerial failure, pricing policies, etc.) has most determined this result. The performance of the intermediate integrated firms, on the

Table 4–1. Market Share by Firm.

	1950			1960			1970			1981		
	A	*B*	*C*	*A*	*B*	*C*	*A*	*B*	*C*	*A*	*B*	*C*
U.S. Steel	22.6	28.4	28.0	18.7	26.3	25.2	21.0	23.1	20.2	16.6	19.1	15.5
Bethlehem	10.9	13.7	13.5	11.4	16.0	15.3	13.8	15.2	13.2	11.6	13.3	10.9
Republic	6.4	8.0	7.9	5.4	7.6	7.3	6.7	7.4	6.4	6.5	7.5	6.1
J&L, Youngstown } merged 1978	6.8	8.5	8.4	7.0	9.8	9.4	8.4	9.3	8.1	7.6	8.7	7.1
National, Granite City } merged 1971	4.0	5.0	5.0	5.3	7.5	7.1	7.3	8.0	7.0	6.6	7.6	6.2
Armco	3.0	3.8	3.7	5.0	7.0	6.7	5.4	5.9	5.2	5.8	6.7	5.4
Inland	3.3	4.1	4.1	5.1	7.2	6.9	4.7	5.2	4.5	5.8	6.7	5.4
Wheeling, Pittsburgh } merged 1968	3.3	2.9	2.9	2.6	3.7	3.5	2.9	3.2	2.8	2.1	2.4	2.0
Total of above	60.3	74.5	73.5	60.5	85.1	81.4	70.2	77.3	67.4	67.6	72.0	58.6
Other domestic	20.3	25.5	25.2	10.6	14.9	14.3	20.6	22.7	19.8	24.4	28.0	22.8

Note: A = millions of net tons shipped;
B = percentage share of shipments by domestic firms;
C = percentage share of shipments by domestic firms + imports.

Sources: *Iron Age*, "Annual Financial Supplement" (various years) and company annual reports.

other hand, has been relatively good. In regard to Inland, National, and Armco this is at least partially due to the fact that these firms' mills are better located than many U.S. Steel plants and are dedicated to what had been dynamic markets in flat-rolled and tubular products. The fact that it is the firms with the most resources that have incurred the greatest losses in market share is one reason there have been relatively few bankruptcies and plant closures—at least up until a few years ago—in spite of low profits and overall decline. Nevertheless, the rate at which large firms—particularly U.S. Steel—are losing market share cannot be sustained indefinitely. This is one reason such firms are diversifying away from steel.

The general contraction of the integrated sector has been accompanied by a quickening pattern of mergers and business failures. However, structural adjustment has been retarded by the traditional oligopolistic structure of the industry and by governmental restraints. No significant horizontal mergers occurred between 1945 and 1968, while at least three have been concluded since that time, presumably a response to the increasing pressure on integrated producers. Typically, however, these consolidations have involved smaller integrated firms that lack the financial resources and scale economies of their larger competitors and that are therefore more vulnerable to the overall pressures weakening the industry. The major steel firms that have been forced into bankruptcy—all since 1979—have been smaller integrated companies (e.g., Alan Wood Steel, Wisconsin Steel, and McLouth Steel). Exit from the industry has been even more pronounced among still smaller firms, many of which have been unable to execute the technological transition away from the open hearth.

Because their large depreciation expenses raise their cash flow, steel firms were sometimes the target of conglomerate acquisitions in the late 1960s. At that time several major companies were acquired by much smaller conglomerates (Jones & Laughlin by Ling-Temco-Vought, Youngstown by Lykes, and Sharon by NVF). Such structural changes do little to improve the competitive prospects of the steel firms involved; there may be some evidence that the reverse is more likely.[4]

Rationalization via consolidation, often associated with some trimming of facilities, is one means by which the steel industry could raise efficiency in a slow-growth market. Yet the largest integrated firms have generally been enjoined from pursuing this option by antitrust considerations. This was the case for the proposed merger between Bethlehem Steel and Youngstown Sheet and Tube, which had been pursued since before World War II. Until the 1960s, Bethlehem's westernmost integrated facility was in Lackawanna, New York; this effectively excluded it from the Chicago-Detroit region, which by the 1950s had become the dominant steel market in the country. The acquisition of Youngstown—a large but relatively weak firm serving the midwestern market—would have brought Bethlehem into this market and by one interpretation would have increased competition in that region by replacing a weak firm with a much stronger one.

In the early 1960s, however, the move was blocked by the government on the grounds that Bethlehem was a potential competitor of Youngstown. Bethlehem responded by entering the midwestern market via construction of what is now its second-largest plant, located at Burns Harbor, Indiana. This resolution has had ambiguous results. On the one hand, were it not for Burns Harbor, the U.S. industry would not have built a greenfield (i.e., entirely new) integrated plant since 1953. On the other, the construction of Burns Harbor contributed to the overexpansion of the industry in flat-rolled products, putting pressure on Youngstown—contributing to its eventual merger with Jones & Laughlin—and suppressing profitability for the industry as a whole.

On balance, it would seem that the interpretation and enforcement of antitrust rules have contributed to the decline of the U.S. steel industry by effectively barring the larger firms from pursuing a more coordinated pattern of rationalization and retrenchment. There is now some evidence that antitrust restraints are being reconsidered. The 1978 merger of Youngstown (Lykes) and Jones & Laughlin (LTV) would almost certainly have been contested by the government had it been proposed even a few years earlier. Youngstown was admittedly moribund, but the merger still involved two of the eight largest firms in the industry and created the industry's third-largest corporation. According to company statements, the merger resulted in economies of scale saving between 85 and 90 million dollars in the first year after the merger—a figure that gives some idea of the potential benefits of such rationalization.[5]

Insofar as the integrated sector has retrenched, this has generally occurred in the historical core of the U.S. steel industry, extending through western Pennsylvania and eastern Ohio to Buffalo, New York. Another traditional steelmaking area, north central Alabama, has also been hard hit. Smaller steelmakers and fabricating plants have been lost in other areas, particularly New England and the West Coast, where the pressure from imports has been most intense. Only the Midwest—especially the Calumet region near Chicago—has been largely exempt from the slow but steady deterioration and loss of facilities. Chicago has long since replaced Pittsburgh as the steelmaking center of the United States—a transition conditioned by the Chicago area's proximity to significant markets. This regional trend has benefited those firms that are centered on that market, for example, Inland and National. Nonetheless, recent developments, particularly the bankruptcy of Detroit's McLouth Steel, indicate that even this region faces decline.

From a long-term perspective this uniform pattern of decline can be linked to the integrated sector's adherence to the performance standards of an oligopoly during a period in which the oligopoly was 'dis-integrating.' This has been most evident in the deteriorating position of U.S. Steel, the former market leader; the ongoing elimination of the U.S. Steel umbrella means the inevitable dismantling of the integrated oligopoly. In spite of this, integrated firms still tend to act as price makers, even though the market power that once permitted them to set prices has eroded.[6] Their organizational structures are often ill equipped to compete aggressively in terms of price and quality, and their traditional methods of cost accounting are not attuned to achieving the lowest possible costs. Caught between nostalgia for the halcyon age of Big Steel and the changing structure of their market, they have generally failed to respond to the contemporary requirements of superior performance.

Diversification is rapidly becoming their preferred response, and this has become the most significant structural change within the integrated sector. This process began in earnest only in the early 1970s. By the end of that decade steel operations accounted for a predominant but no longer overwhelming share of the business operations of firms in the steel industry. From 1979 to 1981, steel operations provided 69 percent of total sales and 59 percent of operating income for the member companies of the American Iron and Steel Institute.[7] The industry leader, U.S. Steel, began to diversify into chemicals in the late 1960s and drastically reduced its dependence on steel through the purchase of Marathon Oil in 1982. In fact, U.S. Steel–almost synonymous with the American steel industry throughout this century–is now best viewed as a diversified oil company. In spite of the sentiment expressed in the quotation that opened this chapter, oil operations now account for a larger portion of the corporation's sales and earnings than does steel. This diversification has had far greater impact than the slow attrition that has weakened U.S. Steel's role in the industry over many decades. While several steel companies–e.g., Bethelehem, Republic, and Inland–have failed to diversify significantly, continued weakness in the steel market may leave them little choice.

THE EMERGENCE OF THE MINI-MILL

Until recently, a discussion of structural changes within the steel industry could have been limited to the themes presented in the preceding section. For all intents and purposes, the integrated sector was the steel industry. Admittedly, the industry has always included a fringe of nonintegrated firms, but these smaller firms traditionally existed only due to the forbearance of their integrated competitors. Such firms accounted for roughly 15 percent of total carbon steel production in the late 1950s.

The share of production claimed by nonintegrated firms in 1981 was not much different. Yet this aggregate constancy masks a complete restructuring of the industry's nonintegrated sector. Through the 1950s, nonintegrated firms were generally small, inefficient, and declining. Lacking the financial resources of their integrated counterparts, they used the same technology—the open hearth—but without extensive economies of scale, without captive sources of raw materials, without extensive marketing networks, and so on.

Today, the nonintegrated sector is made up of small, efficient, and aggressively expanding firms, the mini-mills. They rely primarily on a different technology than their integrated competitors—the electric furnace—and this technology eliminates many of the advantages that formerly accrued to integrated operations. Mini-mills have established more or less complete dominance in several product lines, and in these markets the integrated firms now exist on the forbearance of the mini-mills rather than the reverse. The former fringe of the industry is now the principal locus of the restructuring forces that have been generated by overall secular decline. Far from being a dismissible appendage to the "real" steel industry, the nonintegrated sector has become its dynamic force. In many products, even pricing power (that most cherished desideratum of oligopolistic industries) has shifted to the mini-mill sector.

This is an ironic twist, since one of the sources of mini-mill growth has been the pricing policy of the integrated oligopoly. Substantial price increases on the part of integrated producers during the 1950s (Table 2–9) were one of the factors that attracted entry on the part of mini-mill producers. Yet the success of the mini-mills is not due to the pricing behavior of the integrated sector. The true foundation of mini-mill growth has been the rapid development of advanced and highly appropriate technology. Since the mid-1950s, technical changes have radically altered the traditional structure of scale economies in steelmaking. Whereas in the 1950s nonintegrated firms were puny replicas of their integrated competitors, the technology, performance, and prospects of contemporary mini-mills differ sharply from the integrated norm.

Defining the Mini-mill

The mini-mill sector can be defined along three dimensions: technology, market, and product line. Mini-mills produce carbon steel by melting down scrap in electric arc furnaces, although ore-based inputs may become cost competitive with scrap at some time in the future. This technology eliminates the need for the coke ovens and blast furnaces found in integrated plants. In almost all mini-mills, the steel produced in electric furnaces is then continuously cast into forms suitable for rolling into finished products. This eliminates the need for primary rolling mills. Thus, the standard mini-mill operation—comprising an electric furnace shop, one or more continuous casters, and a rolling mill—incurs significantly lower capital costs than its typical integrated competition.

Mini-mills can also be defined in terms of their market orientation. They have generally been located in smaller regional markets, endowed with local sources of scrap and isolated by transportation costs from competition with other producers or scrap purchasers. Besides their generally local character, mini-mills can also be market defined in terms of product line: they concentrate on relatively simple, low-value commodity products such as wire rod, concrete reinforcing bar, and merchant-quality bars and shapes. This has meant that the more complex products in which economies of scale are significant—especially the flat-rolled sheet used in automobiles, appliances, etc.—have been left to integrated producers, whether domestic or foreign.

There are many firms that do not conform to this description, even though they are clearly part of the mini-mill phenomenon. In recent years especially, several mini-mill firms have outgrown the market limitations that traditionally characterized this sector. Some mini-mill firms (e.g., Nucor, North Star, and Florida Steel) now have a total capacity approaching two million tons—within striking distance of the medium-sized integrated firms. Several recently built or expanded mini-mill plants (e.g., Chapparral in Texas and Raritan in New Jersey) are competing in product lines or quality grades that had long been the exclusive preserve of integrated companies; such plants define their markets as widely as do integrated facilities. The trends described above have rendered the term "mini-mill" somewhat awkward. Largely because it is still the most commonly used term, however, we will use the expression "mini-mill" throughout this book to refer to carbon steel firms that are based exclusively on the electric furnace and have a limited product line. This usage reflects our view that the fundamental engine of mini-mill growth is technological, although this definition does entail the limitation that many integrated firms also maintain plants using similar technology. We will exclude these from our discussion of mini-mills; for a variety of reasons, most integrated producers have had difficulty achieving the same success with electric furnace plants as have their nonintegrated competitors. Finally, we will employ the useful term "market mill" to refer to mini-mill operations that concentrate very strictly on one product, or even a narrow quality or size-range within a product category, but which sell to a dispersed market.

The Growth of the Mini-mill Sector

The history of the mini-mill sector should probably begin with Northwestern Steel and Wire, a producer of simple bar, rod, and wire products located outside Chicago. Northwestern originally produced wire from purchased semifinished inputs. As is often the case with major business innovations, it was necessity—in this case, vulnerability to supply disruptions during the 1930s—that forced the firm to become a steel producer. Due to National Recovery Administration (NRA) prohibitions against integrated capacity expansion, Northwestern adopted electric furnace technology, despite the prevailing wisdom that this was an unprofitable technique for the large-scale production of carbon steel. Northwestern has remained a leader in electric furnace technology, operating some of the largest electric furnaces in the world. By the 1950s, the firm had developed technical refinements that boosted the scale and efficiency of its furnaces tremendously. From 1956 to 1968, for instance, the use of ultra-high voltages and other innovations increased the capacity of one of its furnaces from twenty-one to sixty-three tons per hour.[8] The heat times in moderately sized electric furnaces have decreased from about six hours in the early 1960s to about three hours in the early 1970s and to a lower limit of one hour today. By the late 1950s such improvements had enhanced the attractiveness of the electric furnace to such an extent that it represented a significant market opportunity, one which the mini-mills emerged to exploit.

Rapid adoption of the electric furnace was encouraged by the serendipitous convergence of several other factors as well. Most importantly, the technology for continuously casting billets had become available by the early 1960s; this reduced capital costs and thus lowered the barriers to entry for products made from billets. Furthermore, plants that were built to include this technology enjoyed a significant savings in operating costs due to the elimination of coke ovens and blast furnaces. Second, the integrated transition from the open hearth (which can use a scrap charge of more than 50%) to the basic oxygen furnace (30% maximum scrap charge) increased the availability of economically priced scrap for electric furnace production. Finally, demographic shifts created significant steel markets in the South and West, particularly for the simple products used in construction. These regions were difficult for integrated firms to supply, and they often offered a political and social environment hostile to union activity.

Figure 4–1. Estimated Cumulative Distribution of Steel Production for the U.S. Market, 1961–1981.

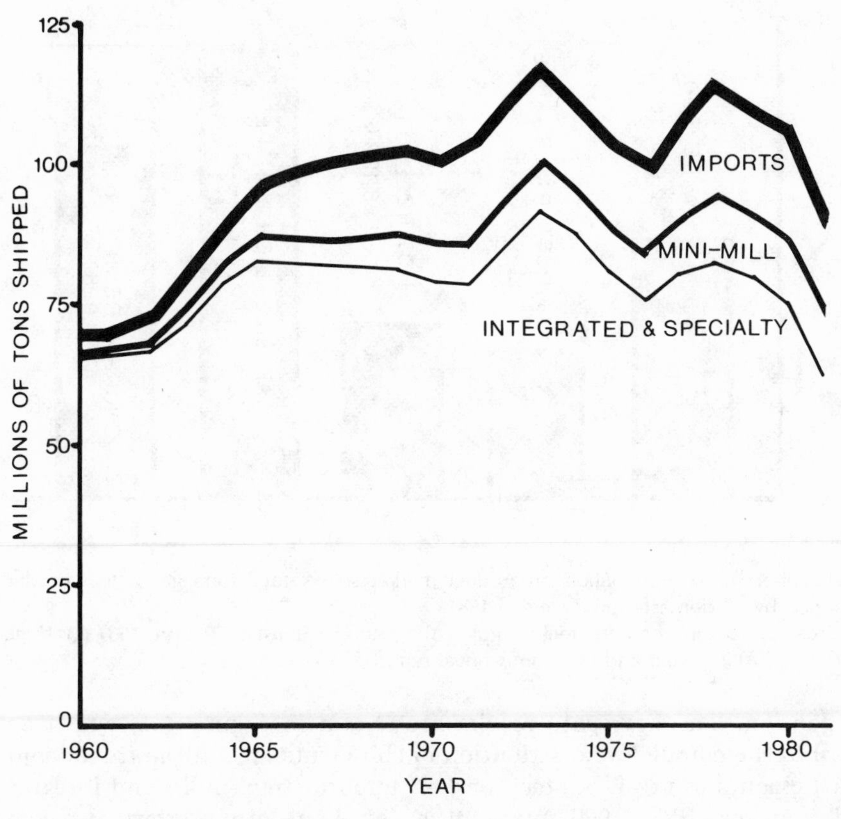

Sources: American Iron and Steel Institute (AISI), *Annual Statistical Report* (New York and Washington, D.C.: AISI, various years) and company annual reports.

The result was the emergence of the mini-mill sector. Many mini-mill firms founded from the late 1950s to the late 1960s developed via backward integration, as steel fabricators recognized the potential savings that could be gained from producing their own inputs. This was the path followed by two of the most successful mini-mill firms: Florida Steel and Nucor; coincidentally, this was also the course taken by Northwestern Steel and Wire in the 1930s. Firms like Florida Steel and Nucor quickly became primarily steel producers rather than fabricators, however, as the success of early experiments led to ever more aggressive

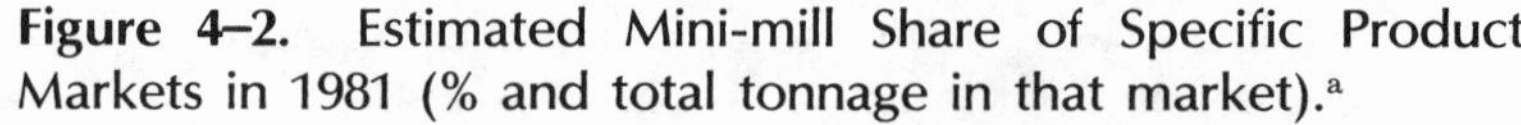

Figure 4–2. Estimated Mini-mill Share of Specific Product Markets in 1981 (% and total tonnage in that market).[a]

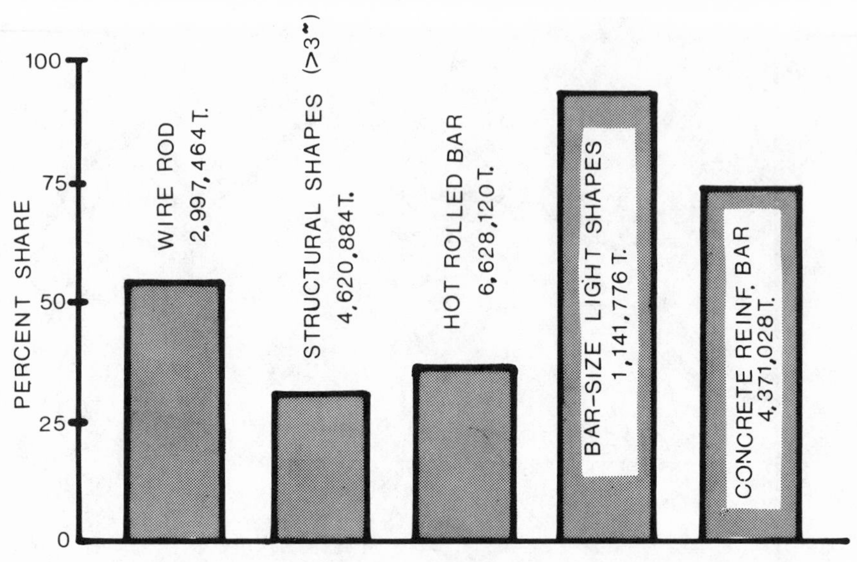

a. Number below designation of product market shows total tonnage of that product shipped by all domestic producers in 1981.
Sources: American Iron and Steel Institute (AISI), *Annual Statistical Report, 1981* (Washington, D.C.: AISI, 1982); and company annual reports.

expansion. The aggregate results are shown in Figure 4–1, which describes the cumulative distribution of shipments among the three major participants in the U.S. steel market: imports, mini-mills, and integrated producers. Since 1960, mini-mill raw steel output has grown at a compound annual rate of almost 10 percent, and market share has grown from less than 3 percent in 1960 to roughly 18 percent in 1982. The most rapid growth in the mini-mill sector has occurred since 1975, when the retrenchment of the integrated sector was at its most intense.

The total mini-mill share of industry shipments does not convey the great strength that these producers have in their markets. At the present time, mini-mill techniques are widely used for only a few commodity-grade products: wire rods, concrete reinforcing bars, bar-size light shapes (e.g., angles), and hot-rolled bars. Since the mid-1960s, and especially during the 1970s, mini-mills have gradually pushed their integrated competitors out of these product lines. Integrated firms have thus either given them up entirely or retreated to the higher quality ranges that have been difficult for mini-mills to produce. Yet the pace of techni-

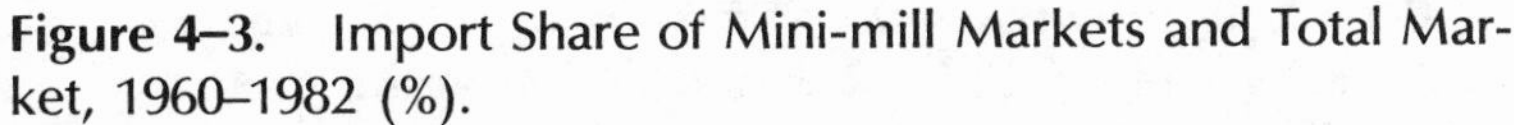

Figure 4–3. Import Share of Mini-mill Markets and Total Market, 1960–1982 (%).

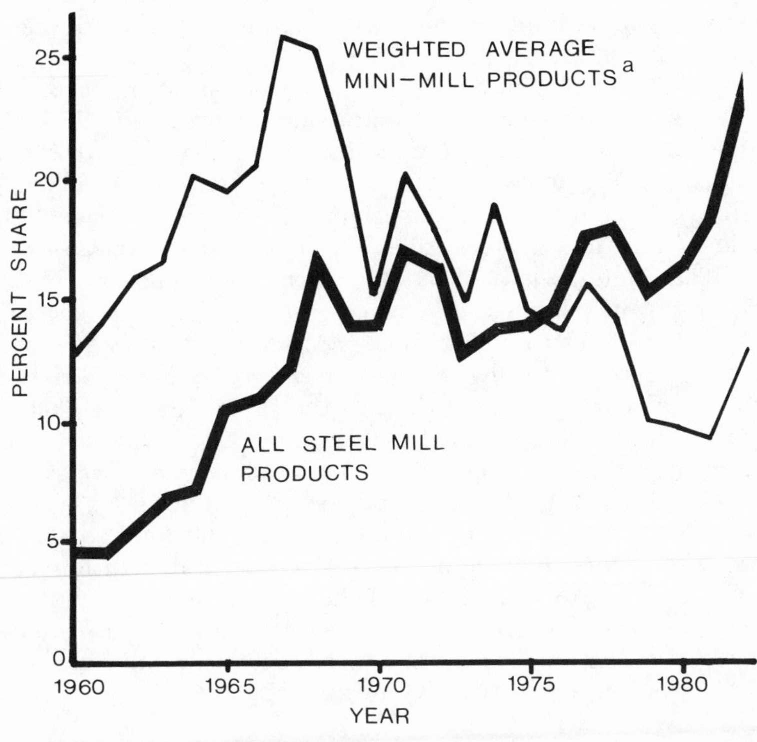

a. Wire rods, light structural shapes, hot-rolled bars, concrete-reinforcing bars; share weighted by total consumption.
Source: American Iron and Steel Institute (AISI), *Annual Statistical Report* (New York and Washington, D.C.: AISI, various years).

cal progress in the mini-mill sector is such that they are already moving into higher quality product lines. Two examples are North Star in bar products and Raritan in wire rods. The eventual elimination of integrated producers from such product categories now seems inevitable, barring fundamental changes in the operating practices of the integrated sector. The authors' estimates of 1981 mini-mill share in various product lines are shown in Figure 4–2.

More surprisingly, perhaps, the mini-mills have enjoyed similar success against foreign producers; this supports the view that the expansion of the mini-mill sector has not been a consequence of poor performance or default by domestic integrated firms. By the end of the 1970s, the mini-mill share of the total market for steel products was roughly comparable to the degree of import penetration, which was 16.3 percent in 1980, but has been much higher since that year. Import penetration is spread across almost all product lines, however; the total mini-mill share is concentrated in only four.

Markets in which mini-mills have a substantial share tend also to have a lower than average import share. More significantly, the trend of imports in mini-mill product lines has contrasted sharply with the overall trend toward an increased import share, at least until 1982 (Figure 4–3). To some extent, this phenomenon may be due to policy-related distortions in the flow of steel trade. This is particularly the case for the period 1969–74 (especially 1969–72), when the Voluntary Restraint Agreements, which limited the tonnage of steel imports, encouraged foreign firms to shift to higher value products; that is, away from the simple merchant-quality products produced by mini-mills. Nonetheless, this cannot account for trends through the 1970s. Nor is it likely that exporting countries simply abandoned these markets in favor of others—an argument that might apply only to Japan and some European countries. More plausible is the view that foreign producers are losing the ability to be profitable competitors in these product lines as mini-mills expand. In other words, U.S. mini-mills have pushed both foreign and domestic integrated producers out of their characteristic product lines. Furthermore, this phenomenon is not restricted to the United States; mini-mill operations have had similar success in other countries—the so-called Bresciani in northern Italy, for example.

Mini-mill Advantages in Technology and Behavior

Spectacular growth has made it possible for mini-mills to enjoy the types of advantages that have accrued to other steel industries in high-growth markets (e.g., the Japanese). Rapid expansion of capacity has been accompanied by continuous technological progress. Each new facility has thus served as a prototype and laboratory for its successors. This sort of technological ferment has produced a steady stream of improvements. Water cooling of the furnace, the use of ultra-high power, oxygen enrichment, ladle metallurgy—all of these advances, pioneered by mini-mill firms, have significantly improved electric furnace technology. Mini-mills have also made dramatic gains through the use of continuous billet casting in lieu of ingot casting and primary rolling. Better sequential pouring, improved water cooling, and other refinements have gradually

Table 4–2. Estimated Average Ages of Steel Facilities in the U.S.: Integrated and Mini-mill.[a]

Facility	*Integrated*		*Mini-mill*	
	Mean Age	*Median Age*	*Mean Age*	*Median Age*
All steelmaking[b]	14.7	15.0	8.1	6.0
Electric furnace	12.1	10.4	8.1	6.0
Primary rolling & continuous casting[b]	25.0	24.5	9.9	6.9
Primary rolling	28.0	27.5	20.5	12.0
Continuous casting	8.0	8.5	6.3	5.2
All rolling mills[b]	21.4	19.4	10.3	6.0
Wire rod	14.3	13.4	11.7	6.0
Bar	33.0	25.0	10.0	6.0
Aggregate[b]	22.0	18.5	9.4	6.3

a. As of January 1, 1982, counting 1981 as year zero.
b. All aggregates are weighted by contribution to total capacity of facilities concerned, with distinct process steps (steelmaking, production of semi-finished shapes, and rolling of finished products) weighted equally.
Sources: 33 Metal Producing Magazine, *World Steel Industry Data Handbook, Vol. 5* (New York: McGraw-Hill, 1982); Institute for Iron and Steel Studies (IISS), *The Steel Industry in Brief: Databook, U.S.A.; 1979-1980* (Green Brook, New Jersey: IISS, 1979); American Iron and Steel Institute (AISI), *Iron and Steel Works Directory of the United States and Canada* (Washington, D.C. and New York: AISI, various years).

lowered casting times, increasing throughput, simplifying operations, and lowering capital costs. Direct rolling of finished shapes, long a goal of steel producers, has been pioneered by mini-mill firms, particularly Nucor.

As is also the case with the Japanese, rapid technical progress and rapid expansion have ensured that mini-mill facilities are generally quite new, suggesting that their technology is correspondingly up to date. The age of facilities in the mini-mill and integrated sectors are compared in Table 4–2, which clearly indicates that the pace of modernization has been radically superior for mini-mill firms. The mean age of facilities is much lower for mini-mill companies, but more illuminating is the pattern of median ages, which are better indicators of the dynamics of modernization. Whereas the mini-mill sector shows a high degree of uniformity in the median age of its equipment, reflecting a consistent process of modernization, the performance of the integrated sector is quite mixed.

The advantages of newer facilities are compounded by the benefits, in terms of capital costs, labor requirements, and so on, provided by the simpler configuration of the standard mini-mill vis-a-vis its integrated rival. Mini-mills are indeed newer, but they are also different; in spite of what the term implies, mini-mills are not miniaturized versions of an integrated plant. Their smaller size does imply that they do not attain the economies of scale associated with large-scale integrated production, but this issue is complicated by the fact that the pattern of economies of scale differs between the steelmaking technologies (see Chapters 6 and 7). At any rate, both their choice of technology and the newness of their facilities provide the mini-mills with a significant advantage over their integrated competitors—so long as attractively priced scrap is available and so long as we restrict the comparison to the characteristic mini-mill products. Superior managerial practices, which could be termed an advantage in "entrepreneurship," represent a third source of mini-mill competitiveness vis-a-vis the integrated sector. While difficult to quantify, this difference has highly significant implications.

Whereas the oligopolistic legacy of price maker status continues to be a burden for the integrated sector, mini-mills behave as price takers. This obviously suggests that mini-mills are more willing to adjust prices in order to maintain sales and that they compete more aggressively in terms of prices. The indirect effects of price taker status may be much more significant, however; in particular, it forces a firm to be much more conscious about reducing costs. Greater cost consciousness on the part of mini-mills is evident on several fronts.

One involves their attitude toward new construction. In contrast to integrated producers, mini-mills adhere to a set of principles that can be summarized by the maxim "build tight, build quick, and build cheap." Building tight involves the elimination of round-out options, redundancies, and crutches such as additional cranes, furnaces, and rolls. The commitment to building quick, which is facilitated by the simplicity of mini-mill technology, results in the possibility of erecting a greenfield mini-mill in less than two years—compared to more than five years for integrated plants. Finally, building cheap implies that unnecessary equipment refinements ("bells and whistles") are avoided and that foundations, infrastructures, and facilities are erected on the premise that the plant may be technologically obsolete in a relatively short time.

Operating costs are also kept to a minimum within the mini-mill sector. Mini-mills generally maintain far leaner headquarters staffs than do their integrated competitors; even the larger mini-mills do not carry significant engineering, market research, or legal staffs. Due to their status as price takers, and because they produce relatively narrow product lines, mini-mills are generally very aware of their cost structure—an indispensable datum for managerial decisionmaking. Integrated firms often have a less thorough grasp of their actual cost structure (e.g., by product); this failing stems from their oligopolistic past, their intricate corporate hierarchies, and their extensive vertical integration. These factors make it difficult for even the best integrated firms to identify their cost structures, to target cost reduction efforts, and to price according to costs.[9]

In a sense, managerial behavior in mini-mill firms resembles what is increasingly being viewed as "the Japanese model," although mini-mill practices are unquestionably indigenous. Both industries have been highly adept at riding the learning curve and developing technological refinements. Both are price takers, so that they are aggressive competitors over market share. Both focus on long-term growth, plowing profits back into facilities.

The parallels between U.S. mini-mills and Japanese firms are most striking, however, in regard to labor practices. Largely because of their regional distribution, U.S. mini-mills are generally nonunionized, while industrial relations in Japan are characterized by relatively pliant company unions. For the mini-mills, the managerial advantages of a nonunion labor force lie not so much in low wages (although basic wage rates are unquestionably lower in mini-mills than in the integrated sector) but in greater flexibility and thus higher productivity. Whereas the structure of work rules is a significant obstacle to productivity growth in the integrated sector, greater efficiency is stressed constantly in the mini-mill environment. Productivity is enhanced by incentive plans that can provide mini-mill employees with income levels comparable to those earned by workers in integrated plants. Productivity gains are also sought by developing a team-oriented corporate culture. Among the more successful mini-mill firms, constant attention is paid to fostering and maintaining a familial, or paternal, relationship with the work force. Even the highest executives are available—at least in principle—to all employees. Internal media are developed both to foster allegiance to the firm and to inform the work force about the economic and competitive prospects of the company. The welfare of all is dependent on the competitiveness of the firm; this message is constantly stressed. Profit-sharing plans, employee stock plans, scholarship programs for dependents, and the like are all employed to develop the same type of corporate culture and identity that is so striking in Japan. In the case of the most dynamic mini-mill firm, Nucor, the parallel with Japanese industrial relations even extends to the principle of lifetime employment.

Nevertheless, there are also crucial differences in the behavior of U.S. mini-mills and Japanese integrated firms. The most telling divergence is in the area of investment strategy. Japanese integrated producers, like their American counterparts, seek to build large, capital-intensive plants in order to gain the maximum economies of scale. A wide range of products—both in type and size—are produced, and the design of plants is fairly complex, in order to minimize the likelihood of contingencies that might constrain output. In contrast, the mini-mill philosophy of "build tight, build quick, and build cheap" leads to smaller plants that have narrow product lines and fewer crutches and redundancies. Furthermore, Japanese integrated firms have built their plants well ahead of the market, so that operating rates are low in periods when demand falters. U.S. mini-mills, by contrast, are typically closely linked to their markets, so that satisfactory operating rates are easier to maintain.

Despite this difference, both the Japanese integrated industry and the U.S. mini-mills have enjoyed comparable success. Mini-mills are highly competitive no matter what the standard by which they are judged: the elegant simplicity of their technical configuration, the newness of their facilities, the aggressiveness of their management, or the flexibility of their industrial relations. As a result, they enjoy a formidable cost advantage over their integrated rivals.

Mini-mill Performance: Costs and Profits

Table 4–3 estimates the 1981 cost of production for wire rods, a representative mini-mill product. The comparisons presented there are both cross-sectoral and international, so that they support several inferences concerning competitive performance. In regard to a cross-sectoral comparison within the United States, Table 4–3 clearly shows the extreme difficulty integrated firms experience in trying to compete in mini-mill product lines. As one would expect given the arguments raised above, mini-mills attain superior performance in every area: raw materials, energy, capital costs, and labor costs. Integrated firms made substantial investments in new wire rod mills in the early 1970s; U.S. Steel, for instance, built a large wire rod mill at its South Chicago Works during this period. Presumably, such investments were undertaken to confront and reverse the expansion of the mini-mills. Table 4–3 shows how hopeless such an effort proved to be. The best mini-mill wire rod mill, Raritan in New Jersey, began operations in 1981 and is targeted directly at the high-quality end of the wire rod market. Its costs are almost certainly below the level described in Table 4–3; it operated near capacity even during the summer of 1982, when the industry as a whole maintained an operating rate below 50 percent and when at least one inte-

Table 4–3. Comparative Production Costs in 1981[a]: Wire Rods (dollars per net ton shipped, normal operating rates[b]).

	Integrated			*Mini-Mill*		
	U.S.	*W. Germany*	*Japan*	*U.S.*	*W. Germany*	*Japan*
Labor	131	84	51	60	45	37
Iron ore	62	50	49	—	—	—
Purchased scrap[c]	15	5	3	93	96	96
Coal or coke	52	75	59	—	—	—
Other energy	46	37	40	45	52	51
Other costs[d]	60	61	64	65	69	68
Operating costs	372	312	266	263	262	252
Depreciation	12	14	16	11	12	11
Interest	5	8	18	7	8	10
Miscellaneous taxes	5	2	4	3	1	2
TOTAL COSTS[e]	393	336	304	284	283	275

a. Exchange rates in 1981, at 2.26 DM/$ and 230 Y/$, were somewhat out of line with historic relationships. As a result, U.S. production costs, especially relative to those estimated for Europe, may be slightly overstated.
b. For integrated plants, average 1977 to 1981 capacity utilization rates were used to avoid single-year abnormalities. For the United States, capacity utilization averaged 80 percent, for West Germany and Japan, 65 percent. For mini-mills, 85 percent capacity utilization was assumed throughout, and this is close to their average over the last five years.
c. Average 1980–81 scrap prices were used, since these are more typical of long-term relationships than 1981 scrap prices alone.
d. Includes alloying agents, fluxes, refractories, rolls, and so on.
e. Excluding any return on equity.
Sources: Estimated by authors from data contained in annual reports (e.g., Korfstahl, Tokyo Steel, Florida Steel, etc.), Metal Bulletin, World Steel Dynamics, *Core Report Q* (New York: Paine Webber Mitchell Hutchins, 1982), and so on. See Appendix C.

grated wire rod mill built during the 1970s, the one at U.S. Steel's South Chicago Works, was completely shut down.

Table 4–3 also presents estimates of wire rod costs in Japan and West Germany. The international competitiveness of U.S. mini-mills is evident, as is the lack of competitiveness on the part of U.S. integrated producers. Japan's cost advantage in the integrated production of wire rods is even greater than is the case for cold-rolled sheet (see Table 3–6). Given the now familiar Japanese cost advantage in integrated techniques, the competitiveness of U.S. mini-mills may be surprising. In fact, Table 4–3 reflects the fact that the U.S. steel industry has been broken into two distinct parts; judgments that apply to the still dominant integrated sector need not, and probably will not, hold true for the mini-mill

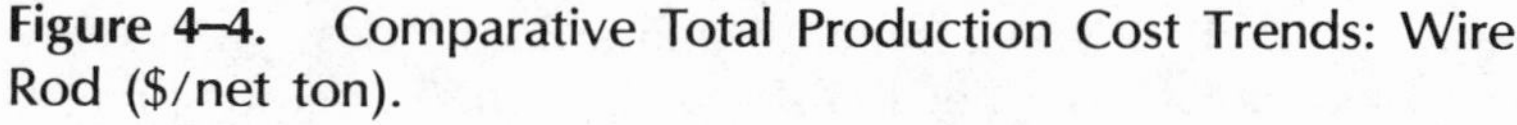

Figure 4–4. Comparative Total Production Cost Trends: Wire Rod ($/net ton).

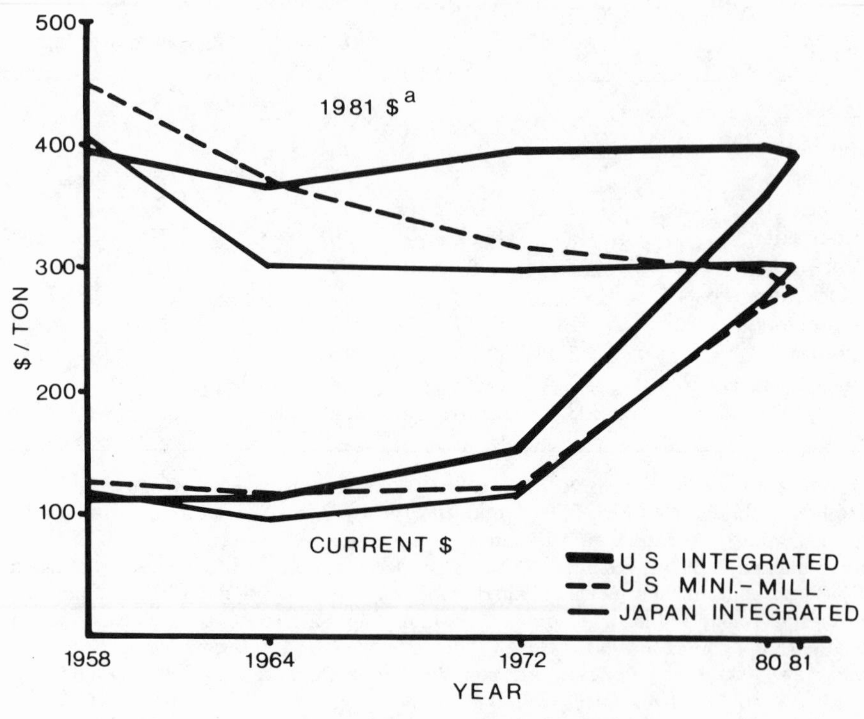

a. Converted from current to 1981 dollars using the Producer Price Index for Industrial Products of the U.S. Department of Labor, Bureau of Labor Statistics.
Source: As in Table 4–3.

sector. The outstanding performance of U.S. mini-mills, even when judged against the Japanese, is evidence that the U.S. steel industry is not inherently uncompetitive.

Finally, Table 4–3 suggests that superior integrated practices cannot offset the mini-mill advantage. Even the most efficient integrated producer, the Japanese, cannot match mini-mill costs in the production of wire rods. The mini-mill advantage is understandably greatest where integrated performance is poor and where scrap prices are low (the U.S.) and smallest where integrated performance is strong and scrap prices high (Japan). In the latter case, scrap price increases could eliminate the mini-mill advantage. Nevertheless, the data suggest that in wire rods even superior integrated performance no longer guarantees competitive-

Table 4–4. Return on Equity, Mini-mill versus Integrated (yearly average, %).

	1972–76	*1977–81*
Mini-mill composite[a]	15.8	17.0
Integrated composite[b]	9.1	5.6
Total industry	9.4	7.2

a. Nucor, Northwestern Steel and Wire, Florida Steel and Cascade Steel Rolling Mills, weighted by total sales.
b. U.S. Steel, Bethlehem Steel, Inland Steel, and Republic Steel, weighted by raw steel production.
Sources: American Iron and Steel Institute (AISI), *Annual Statistical Report, 1981* (Washington, D.C.: AISI, 1981); and company annual reports.

ness against mini-mill techniques. Cost trends have favored mini-mill producers since the mid-1970s; this is illustrated in Figure 4–4, which describes historical trends in wire-rod costs for U.S. integrated producers, Japanese integrated producers, and U.S. mini-mills.[10] The cost trends described in this graph, which is the wire rod analog of Figure 3–6, suggest that mini-mills have been very successful at controlling costs, in spite of an early disadvantage in the 1950s, while integrated producers have achieved much less success. In the United States, mini-mills had achieved parity with integrated producers by the mid-1960s, and their advantage has been intensified by the macroeconomic changes referred to at the beginning of this chapter. Mini-mill technology—based on the principles of saving labor and reducing capital costs—is a highly appropriate response to the forces which are altering the U.S. and world economies.

Lower costs generate higher profits, as is shown in Table 4–4, which compares profitability for domestic integrated and mini-mill firms. The data presented in Table 4–4 are derived for a sample set of firms from each sector.[11] The striking discrepancy in the rate of return on equity between the mini-mill and integrated sectors provides the principal immediate explanation for the rapid growth of mini-mill firms. Higher mini-mill profits have encouraged the flow of capital into this sector, feeding its rapid expansion. This in a sense closes the circle of mini-mill competitiveness: high profit rates attract investment, which improves technology, thus reducing costs and in turn boosting profits—a virtuous circle that contrasts sharply with the integrated pattern.

While the prospect of higher profits has attracted capital to the mini-mill sector, this has rarely come from integrated firms, which have generally failed to exploit the advantages offered by mini-mill technology. One can only conjecture about the reasons for this. Labor costs certainly play a role; unionized integrated producers would find it difficult to establish nonunionized, mini-mill type plants. Employment costs are not the principal source of the mini-mill advantage, however. It is highly likely that integrated firms simply failed to grasp the fundamental changes connected with the rise of the mini-mill sector. Accustomed to assuming their own competitive superiority and the appropriateness of their technology, integrated producers—especially at the highest executive level—have long dismissed nonintegrated producers as a peripheral element in the steel industry. Integrated producers are ill suited to adopt the flexible and aggressive managerial style so evident in the mini-mill firms.

Foreign capital has not failed to grasp the potential of mini-mill production in the United States. Several of the most successful mini-mills (e.g., Georgetown, Raritan, Bayou) are wholly owned by foreign firms; others have substantial foreign participation (e.g., Chapparral, which is 50% Canadian owned). Foreign capital has been especially active in expanding the frontiers of the mini-mill sector, for example, in the construction of market mills. One of the more path-breaking market mills now being planned, York-Hannover, is Canadian owned. This mill will produce seamless tubes, representing mini-mill entry into yet another product line that has until now been controlled by integrated firms.[12] In sharp contrast to the lack of international investment in integrated steel production (except for limited regions of Western Europe), mini-mill production represents a highly multinational segment of total world steel output.

Prospects

Growth in the mini-mill sector has been so rapid that these firms are now graduating to a new role in the U.S. steel market, with new problems and prospects. Once seemingly recessionproof, most mini-mill firms were hit hard by the recession of 1982—a fact that might suggest that the mini-mill phenomenon has natural limits.

One potential barrier to further mini-mill expansion concerns the availability and quality of scrap, the principal input to mini-mill operations. Largely due to the expansion of the mini-mill sector, some lessons of which are increasingly being drawn by integrated firms as well, the electric furnace share of U.S. raw steel production has increased at a compound annual rate of almost 5 percent since 1971. In 1981, the electric furnace share of total steel output was 28.3 percent, up from 17.4

percent in 1971.[13] In spite of this increase, however, the real price of scrap has not risen.[14] It is true that scrap prices are volatile, and there have been periods (e.g., 1974) when they were relatively high. High scrap prices will be associated with high steel prices, however, and this dampens their effects on profits.

There is no evidence of a secular rise in scrap prices, and scrap quality has yet to limit the continued expansion of electric furnace operations. Even were such problems to arise, however, there is little likelihood that they could halt the advances being made by the mini-mill sector. High scrap prices would create a market opportunity for the development of more sophisticated retrieval techniques or an alternative to the electric furnace. Oil- or gas-based direct reduction techniques are already in commercial operation, particularly where energy costs are low. At present, it costs roughly $130 to produce a ton of directly reduced iron, so that it is uncompetitive with scrap. Even were mini-mills forced to rely on directly reduced iron at that price, however, it would still leave them with a significant cost advantage against integrated producers. Scrap prices could more than double from the level assumed in Table 4–3 and still leave the mini-mills with a cost advantage. Alternative inputs are available at prices below this level, so that the mini-mill advantage is secure—especially since they are improving efficiency more rapidly than their integrated competitors. Moreover, research into alternatives to scrap is still at an early stage; coal-based direct reduction or other processes, such as direct plasma steelmaking, will eventually lower the relative price of ore-based inputs to electric furnaces.[15] All of these arguments suggest that scrap supplies are unlikely to be a significant constraint on mini-mill expansion, especially since increasing scrap prices have yet to prove a problem except for short periods. In late 1982, the actual scrap price was, at $50 per ton, so depressed due to the recession that it was only 60 percent of the level used to construct Table 4–3.[16]

A more plausible constraint concerns the range of products for which mini-mill techniques are suitable, although here as well the barriers are far from insurmountable. As mini-mills drive integrated producers out of the markets for commodity-grade rods and bar products, continued growth becomes more problematic. There is little doubt, however, about the ability of the more successful mini-mill producers to move gradually into markets that are still dominated by integrated firms, especially higher quality bar and rod products (e.g., cold finished), heavier structural shapes, and some flat-rolled products. Mini-mill firms have already successfully penetrated all of these markets but the last. While flat-rolled markets are often viewed as the inviolable bastion of integrated techniques, even this is a misconception. Mini-mills are already potential producers of some flat-rolled products (e.g., narrow strip). Substantial mini-mill penetration of these markets must await the development of a technology capable of casting thin slabs—an innovation that would simplify the process of rolling sheet from slabs and thus reduce the impact of scale economies in these product lines. Once this is achieved, mini-mills may be highly competitive in narrower and less sophisticated flat-rolled products, which make up a significant share of the overall flat-rolled market. Mini-mill entry into a substantial segment of the flat-rolled market would alter the entire cost structure for this product category by forcing integrated producers to concentrate on high-grade and wider flat-rolled products, reducing production runs and scale economies.

Thus, there is no reason to suspect that the mini-mill sector has reached its limit. It is also true, however, that further structural changes in the industry will tend to blur the distinction between the two sectors, as mini-mills penetrate more markets and as integrated firms adopt mini-mill techniques. Mini-mill adoption of direct reduction or plasma steelmaking techniques will represent their backward integration (some have already purchased scrap suppliers), knitting the two sectors further together. But it is the traditional sector that will be forced to adapt to the mini-mills rather than the reverse.

CONCLUSION

For members of the steel community, there is an understandable tendency to view the rivalry between the integrated and mini-mill sectors from within the horizon defined by the industry itself. Yet the recent changes within the industry—especially the spectacular emergence of the mini-mills—are primarily reflections of momentous structural changes occurring within the U.S. economy as a whole. As the economy changes, so does the place of steel within it—a fact shown most clearly by the trend in steel intensity. The structural changes described in this chapter define the process through which the steel industry is adapting to a radically different economic environment.

We began this chapter by listing some of the broader forces that are reshaping the American, and world, economy and which together define an industrial revolution. Their most salient feature is the widespread adoption of advanced electronics. Technological developments like computers, word processors, and industrial robots promise to redefine the nature and requirements of the work process, promoting the fragmentation of production and the dispersal or deurbanization of population.

This process represents a contemporary redefinition and refinement of the division of labor, since it is likely to involve a breaking up of the massive factories that characterized the last industrial revolution. Mini-mills are one reflection of this trend.

The American steel industry has been undergoing a process of fragmentation since at least the 1960s. Even before then, the integrated oligopoly began to lose control of various markets—e.g., wire products, which constituted over 6 percent of total industry shipments in the early 1950s but which are for all practical purposes no longer produced by steelmaking firms in the United States. Increased market penetration by mini-mills has also diluted the concentration of U.S. steel production. A further example of this process is the growth of the steel service centers, a separate industry which distributes steel mill products to consuming firms. Service centers are increasingly expanding beyond a warehousing role and carrying out some productive functions, such as cutting to shape. More steel is now shipped through service centers than to the automobile industry, making service centers the largest purchaser of steel mill products. Service center growth is likely to continue, given the probability of increasing import penetration, mini-mill expansion, and the efforts of consuming industries to reduce inventories in response to high real interest rates. Efficiency has increased as a result of this fragmentation of functions once dominated by integrated steelmakers.

In contrast with their integrated competitors, mini-mills are well positioned to benefit from the forces that are restructuring the American economy. Mini-mills are much less energy intensive than their integrated rivals, and this suggests their greater appropriateness for an era in which the opportunity cost of energy is likely to remain high. More significantly, mini-mills are likely to benefit from increasing real capital costs. Historically high debt levels and the imperative of maintaining technological competitiveness suggest that high real interest rates may persist. This will increase the advantage represented by the relatively low capital costs of mini-mill plants, encouraging the further expansion of this sector.

These points suggest the context in which the emergence of the mini-mill sector assumes its true significance. Masked by the common perception of the steel industry as stagnating or declining, the real picture is one of fundamental change, both in market structure and in technology. The days when the integrated sector could be identified with the entire industry have passed. Over the past fifteen years, the appropriate focus for analyzing the dynamics of the industry has been not the integrated sector but its mini-mill competitor and the interface between these two. This will be even more obvious in the future.

Notes

1. "U.S. Steel: Gambling Heavily on the Petrochemical Business," *Business Week*, October 9, 1978, p. 70.
2. Daniel Seligman, "The Year There Was No Inflation," *Fortune*, May 5, 1980, p. 100.
3. U.S. Steel Corporation, "Annual Report" (various years) and American Iron and Steel Institute (AISI), *Annual Statistical Report* (New York and Washington, D.C.: AISI, various years).
4. Barry Bluestone and Bennett Harrison, *The Deindustrialization of America* (New York: Basic Books, 1982), p. 152.
5. "LTV: On the Acquisition Trail Again, but Now in Aerospace and Energy," *Business Week*, November 3, 1980, p. 66.
6. "Steel Pricing That Ignores the Market," *Business Week*, February 2, 1981, p. 26.
7. AISI, *Annual Statistical Report, 1981* (New York, 1982), p. 13a.
8. Joseph C. Wyman, "Steel Mini-Mills—an Investment Opportunity?" (New York: Shearson Loeb Rhoades, Inc., 1980), p. 25.
9. Michael Rosenbaum, "Inland Steel Launches Mini-Mills Counter-offensive," *Journal of Commerce*, November 1, 1982, p. 5a describes how mini-mill competition has forced this highly efficient integrated producer to alter its pricing structure. Formerly, this firm followed a pricing strategy that only loosely linked prices and costs.
10. It should be noted that this estimate of 1958 mini-mill costs is somewhat hypothetical, since this sector emerged around 1960.
11. Several factors—principally diversification and the large number of privately held mini-mills—make it difficult to calculate comprehensive measures of mini-mill versus integrated profitability, so that it is necessary to use a representative sample. The mini-mill composite includes three of the largest and most successful mini-mill firms—Northwestern Steel and Wire, Nucor, and Florida Steel—as well as Cascade Steel Rolling Mills, a smaller mini-mill located in Oregon. These firms, which are exclusively engaged in steel production or fabrication, were weighted according to their net sales in order to develop the mini-mill composite rate of return on equity. U.S. Steel, Bethlehem, Republic, and Inland constitute the integrated composite. The latter three firms are only slightly diversified; U.S. Steel, in spite of substantial nonsteel operations, is included in this sample because of its status as the industry leader. Due to the diversification of U.S. Steel, the profitability of each integrated firm was weighted by its net steel shipments to develop the composite for that sector. The inclusion of nonsteel income in the integrated sample will tend to overstate the rate of return for

that sector's steel operations. In spite of this, however, comparison of these results with the industry's average return on equity (also shown in Table 4–4), suggests that for the 1977–81 period the integrated sector's profit rate must be weighted by 85 percent (vs. 15% for the mini-mill rate) to derive the industry average. Since this is roughly comparable to the integrated share of industry shipments for this period, it provides indirect evidence for the accuracy of these composite estimates.

12. See, for example, G.J. McManus, "Now the Slump Threatens to Pinch the Mini-mill Growth," *Iron Age*, August 2, 1982, p. MP-7.
13. AISI, *Annual Statistical Report, 1981*, p. 55.
14. *1982 Metal Statistics* (New York: American Metal Market, 1982), p. 192 and *Economic Report of the President* (Washington, D.C.: U.S. Government Printing Office, 1982), p. 301.
15. See, for example, Robert R. Irving, "The Inred Process: an Ironmaking Option," *Iron Age*, July 6, 1981. Hot metal produced by this process is estimated to have variable costs roughly 60 percent of those needed for blast furnace production.
16. "Weekly Steel Scrap Price Composite," *American Metal Market* (various issues).

II THE LESSONS OF PRODUCTIVITY PERFORMANCE

5 THE PRODUCTIVITY MATRIX: PRODUCT BY PROCESS

> *The next "scientific revolution," the overturning of the paradigms that underlay economic theory and economic policy these last thirty years, may start with productivity or with capital formation.*[1]
>
> – *Peter F. Drucker*

Part I of this book has discussed the postwar history of the U.S. steel industry in the light of international standards of performance. The image that emerges from this description is an uneven one. One area, the growing mini-mill sector, has shown outstanding performance. The record of the dominant integrated sector, however, is mixed. In spite of substantial investment and the partial adoption of new technologies, profits have been consistently low, market share has shrunk, and capacity has contracted. Foreign producers, especially the Japanese, have been more dynamic and successful participants in the world steel industry. U.S. firms can justifiably claim superior performance vis-a-vis some of their European competitors, but this is less an indication of American success than a reflection of the declining competitiveness of traditional steelmaking regions and technologies.

The historical description that was presented in Part I is broad enough to indicate the underlying forces that have determined the secular decline of the American steel industry's integrated sector. The present difficulties of this sector are not due to faulty policies or cyclical pressures; historical investigation shows that the problems have much deeper roots. Yet the treatment in Part I suffers from the limitations of all such historical analysis: it is broad enough to indicate the existence of structural change but not specific enough to assist managers and policymakers in coping with the developments that structural change portends. Part II is designed to rectify this by presenting a highly specific analysis of how the American steel industry lost its productivity advantage against the Japanese. This chapter introduces this theme by providing a detailed microeconomic comparison of labor productivity in the U.S. and Japanese steel industries since 1958, looking at the record of productivity growth by process and by product. Chapter 6 will then analyze the data presented in Chapter 5, discussing the various sources of comparative productivity performance. Chapter 7 will extend this analysis to consider not just labor productivity but overall costs (especially capital costs) and on that basis suggest ways in which the future performance of the U.S. industry could be improved.

THE IMPORTANCE OF LABOR PRODUCTIVITY

In the broadest sense, productivity is a synonym for physical efficiency and is thus the fundamental determinant of performance. Whereas efficiency in the market is measured in terms of costs, the physical measure of efficiency is factor productivity, that is, output per unit of input. Overall trends in productivity can be described in terms of total factor productivity, the residual growth of output beyond increases in the quantity of inputs. Productivity can also be specified in terms of the inputs to the production process: one can consider output per unit of labor time (labor productivity), per unit of capital (capital productivity), per unit of energy (energy efficiency), or per unit of any raw material. All of these measures contribute to the overall evaluation of economic efficiency in any society, industry, or firm. As economic conditions change, certain measures may assume greater importance; this has been the case for energy efficiency, for instance, since 1973. Nonetheless, labor productivity is generally recognized as the most illuminating physical measure of efficiency.

This has been especially true in recent years, when the competitive problems of several American industries have focused increased attention on trends in labor productivity. The erosion of the postwar U.S. advantage in labor productivity seems to define the declining prospects of many American industries; employment costs which have traditionally been high by international standards make labor productivity a crucial variable for U.S. firms. Furthermore, lagging productivity growth is linked to the seemingly intractable symptoms of stagflation: deficits in the balance of trade, persistent unemployment, slow growth in real incomes, and so on. Among economists, stagflation has battered the Keynesian orthodoxy and has elicited a vigorous search for an alternative paradigm, one that emphasizes the sphere of production (the supply side) rather than the categories of aggregate demand which dominate the Keynesian approach. Within this context, the broad synthesizing potential of labor productivity has ensured that it will play a central conceptual role in the development of any supply-oriented rival (or complement) to Keynesian economics.

One reason for the increased interest in labor productivity is the topical importance of the problems generated by lagging productivity growth. Table 5–1 shows how U.S. labor productivity grew very slowly through most of the 1970s, both in the steel industry and in the economy as a whole. Declining rates of productivity improvement have slowed GNP growth. Higher rates of productivity growth in previous periods had generated increased real incomes, increased demand for goods and services, and thus increased economic activity. During the 1970s and 1980s, however, slow productivity growth has disrupted this pattern. Output per hour worked has grown slowly, and real incomes have stagnated. Furthermore, slow growth in output per unit of labor time has ensured that increases in employment compensation are less likely to be offset by improved efficiency; as a result, increased costs rapidly lead to increased prices and thus perpetuate an inflationary spiral. For these reasons, trends in labor productivity are at the core of the macroeconomic problems that have weakened U.S. economic performance over the past ten to fifteen years.

The importance of labor productivity should not be reduced to topical relevance, however. Its theoretical significance is equally weighty, largely because of its synthesizing potential in describing supply-side performance. Labor productivity is a prime determinant of performance in its own right, since when combined with employment costs it defines unit labor costs. Just as importantly, labor productivity is often an appropriate proxy for the rate of technical progress, the efficiency of investment, and the productivity of other inputs. Capital productivity may be equally important, but the direct evaluation of capital productivity encounters difficulties connected to the measurement of capital as a physical input. While not completely exempt from such problems, labor productivity is a much more readily applied concept. Other measures of efficiency—in terms of energy, raw materials, and so on—tend to be closely correlated with labor productivity, so that insights into trends

Table 5–1. Comparative Annual Productivity Changes (%).

	Steel	*Nonfarm Business*
1951–58	1.3	1.9
1958–64	2.6	3.0
1964–72	2.0	2.2
1972–80	1.1	1.0

Source: U.S. Department of Labor, Bureau of Labor Statistics.

in labor usage will frequently apply to other inputs as well. All of these considerations make labor productivity the most useful measure of overall physical efficiency. Most important, perhaps, is the philosophical relevance of this concept; concentration on labor productivity reflects the recognition that production is primarily a social process, the final purpose of which is improving the welfare of human beings.

The most frequently used measures of labor productivity are macroeconomic indices of value product, such as those listed in Table 5–1. Such aggregate measures of labor productivity can be applied to the economy as a whole or to a specific industry, and we will refer to this type of productivity analysis as the macro approach. The trends described by such indices are generally explained as the result of technical changes and the substitution of capital for labor, which boosts productivity by raising the capital-labor ratio. Technical change is often treated as a residual in quantitative analyses of the relationship between labor productivity and capital formation.[2] This traditional theoretical interpretation of the forces that determine productivity growth can be applied to explain the recent drop in productivity growth from its historical norm of about 2.5 percent per year to about 1 percent per year in the 1970s. Reduced rates of productivity improvement may be ascribed to inadequate capital formation and savings. Poor productivity performance may also be linked to a lack of significant new technologies or to a general failure to implement available innovations. Both these interpretations are based on the macroeconomic analysis of productivity growth, and both entail reasonable policy implications: productivity performance will be improved by increased incentives for saving and investment, greater commitment to research and development, and so on.

The same factors can be cited to explain the slowdown in labor productivity in the steel industry, which mirrors the pattern evident in the economy as a whole. Investment in the steel industry has been declining in real terms, so that the industry's capital-labor ratio has been basically stagnant. Furthermore, the industry's record of technological progress in the 1970s has been lacklustre. This would tend to confirm the relevance of the macroeconomic theory of labor productivity for analyzing trends in the steel industry's performance. This type of productivity analysis, applied to an industry as a whole, can be a useful guide to evaluating overall performance and government policies.

The Micro Approach to Productivity

Unfortunately, the broadness of the macro approach to productivity limits its practical relevance. Macro measures can be cited to support general policy initiatives or to generate after-the-fact descriptions of developments within an industry. For the steel industry, for instance, industrywide productivity comparisons show that a once substantial U.S. lead has been eroded or, at least in the case of Japan, erased. Paradoxically, however, such comparisons have little or no operative significance for improving performance. Although macro measures strikingly illustrate the urgency of the industry's situation, they are too general to have any direct impact on firm behavior. Knowing the position of the U.S. industry as a whole relative to its international rivals tells an individual steel firm little about its own performance.

Even were a firm able to exactly rank its own aggregate performance relative to international standards, it would still be unable to define specifically its own advantages or disadvantages and to identify the most promising areas for investment. At the decisionmaking level of the firm, less esoteric and more concrete measures of productivity are essential. This implies the adoption of a micro approach to productivity analysis, one that disaggregates the production process into discrete steps in terms of which labor productivity can be evaluated for various products. Macroeconomic productivity measures are useful; against the background of contemporary stagflation, they deal with the right theme. Nevertheless, they retain an increasingly awkward aggregate framework, one that is ill-suited to addressing contemporary economic problems.

The availability of microeconomic productivity data is a little known but nonetheless significant factor in the outstanding postwar performance of the Japanese steel industry. In 1958, Japanese steel producers required 35.7 manhours per net ton (MHPT) for the production of cold-rolled sheet, the product for which most of our integrated comparisons will be made; this had been reduced to 5.8 MHPT by 1980. While there are several reasons for this high level of performance, it has certainly been facilitated by the fact that the Japanese government maintains detailed data on the actual MHPT requirements for each individual process employed in the production of each of the major products produced in a steel mill.[3] These data, which are available to Japanese firms on an industrywide basis, can potentially eliminate firms' uncertainty about their competitive standing in regard to labor productivity. This has probably increased the "x-efficiency" of Japanese firms—that is, the extent to which they attain the maximum output possible given the inputs they use.[4] Without a knowledge of what has been achieved by competitors, the potential for increasing x-efficiency is sharply reduced. Finally, these data enable Japanese firms to identify which processes hold the greatest potential for improving productivity. The availability of process-productivity statistics thus reduces the uncertainty associated with investment.

The maintenance of such a data base in Japan reflects the Japanese commitment to improving labor productivity; Japanese firms have used such data to evaluate performance from year to year and to discover ways to make ever greater progress in individual processes. The Japanese data can also be used to evaluate the potential of substituting one process for another. Among U.S. firms, however, no comparative process data have been kept on a continuing basis, and little attempt has been made to evaluate and improve performance by applying such data. Insofar as American companies concentrate on historical improvements in labor productivity, their records are usually kept in terms of total labor requirements by product rather than by process. Even the detailed process productivity data available for Japan, however, have not been linked to the production of specific products. A truly microeconomic description of labor productivity in the steel industry would ideally combine both elements, process data and product data. This requires some means of aggregating process data in order to devise product totals while "revealing" the importance of various processes to the final outcome. The missing link for the combination of process and product information is an input-output table that describes yield relationships as production moves from step to step. The contribution of the authors has been to combine the process productivity data for key operating steps with such an input-output table, so that a detailed cumulative representation of labor requirements (MHPT) can be described for each product. The result could be conceived as a matrix describing labor requirements for each process in the production of each product. Once this description of productivity relationships has been developed for a historical period, it is then possible to delineate trends by process and by product. On this basis, finally, it becomes possible to identify and analyze, with a high degree of accuracy, the factors that determine productivity performance.

Productivity by Process and by Product: Methodology

While the principles which underlie the analysis of productivity by process and by product are relatively easy to describe, the actual construction of this data base has been an arduous process. The data sources, assumptions, and other methodological issues used in developing and interpreting these productivity data are discussed in Appendix C.

One of the problems inherent in a microeconomic treatment of productivity is the wealth of detailed data upon which such an analysis must be based. This makes it inappropriate to reproduce all of the results of this research in the format of this book. Although data comparing U.S. and Japanese productivity have been collected on fifteen products and 20 processes processes for most of the years between 1958 and 1980, only a representative sample will be presented here. In most cases, the discussion will be restricted to performance in 1958, 1964, 1972, and 1980—years chosen not only because they are at more or less regular intervals but also because they are characterized by comparable operating rates. Only major product categories will be discussed. Special attention will be given to cold-rolled sheet among integrated products and to wire rod, which is produced by both the integrated and mini-mill sectors of the industry.

Throughout this analysis, an effort has been made to treat these two sectors as distinct, since this allows one to grasp the productivity effects of ongoing structural changes within the world steel industry. Comparisons are made on the basis of the average or representative characteristics of the two countries and the two sectors—regardless of whether these characteristics are actually embodied in any existing plant. In general, productivity tables have been constructed as if the operating parameters of the entire sector, in each country, were found in a single plant. This may be misleading. Integrated mills, for instance, have begun to adopt what we will characterize as mini-mill techniques (electric furnaces combined with continuous casters) for the production of barlike products such as wire rods, so that the depiction of integrated MHPT for wire rod production may be somewhat hypothetical. The integrated configuration that is described for wire rod production is in fact disappearing as integrated producers set up facilities modeled in part after mini-mill shops, but this trend confirms rather than invalidates the data presented here.

The end result of the process of data collection and interpretation is a series of detailed annual product tables, describing the cumulative process requirements by product for industries and sectors. One example of such a table is provided by Table 5–2,[5] which describes the detailed disaggregation of labor productivity for the production of cold-rolled sheet in Japan and the U.S. in 1980. The following procedures have been used to construct both this table and those for other years and other products that are not shown:

- For each production process, MHPT requirements for the output *of that process* are shown in the left column, the input-output relationship between each process step is shown in the center, and cumulative MHPT (through all previous process steps) are presented in the columns at the right. For example, in Table 5–2, U.S. total cumulative MHPT through the coke oven stage (0.87) is calculated by add-

Table 5–2. Comparative Manhours per Net Ton by Process: Cold-rolled Sheet, 1980 (actual operating rates).

	Japan Process Direct (detail)	Japan Process Direct	Japan Process Total (detail)	Japan Process Total	Japan I/O	Japan Cumulative Direct	Japan Cumulative Total	United States Process Direct (detail)	United States Process Direct	United States Process Total (detail)	United States Process Total	United States I/O	United States Cumulative Direct	United States Cumulative Total
Ore handling				.04							.06			
Sintering		.02		.16					.05		.25			
Coal handling				.04							.05			
Coke ovens		.31		.53		.31	.60		.48		.80		.48	.87
Ore/HM					1.55							1.49		
Coal/HM					.76							.84		
Sinter/HM					1.27							.36		
Coke/HM					.52							.57		
Blast furnace		.10		.34		.29	.93		.27		.65		.56	1.33
HM/CS					1.00							.80		
Scrap/CS					.09							.36		
Steel furnaces														
Melt:	(1.00) .14		.41					(1.00) .32		.57				
OH	(0.00) -		-					(0.16) .60		1.10				
BOF	(1.00) .14		.41					(0.84) .27		.47				
I.C.	(0.42) .21	.30	.25	.66		.59	1.59	(0.80) .11	.48	.31	.96		.93	2.02
OH	(0.00) -		-					(0.16) .18		.52				
BOF	(1.00) .21		.25					(0.84) .09		.26				
C.C.	(0.58) .13		.25					(0.20) .36		.72				
CS/Slab					1.060							1.11		
Slab	(0.42) .36	.15	.68	.29		.78	1.98	(0.80) .34	.27	.63	.50		1.30	2.75
Slab/HR					1.048							1.09		
Hot mill	.18		.47					.24		.53				
Condition, trim	.12	.30	.32	.79		1.11	2.87	.18	.42	.40	.93		1.84	3.93
HR/Fin. CR					1.050							1.10		
Pickle	.14		.19					.17		.22				
Cold reduction	.16		.22					.16		.22				
Anneal	.16	.98	.22	1.33		2.15	4.34	.20	1.22	.24	1.57		3.24	5.89
Temper	.18		.24					.27		.34				
Finish	.34		.46					.42		.55				
Package & shipping				.88			5.22				.78			6.67
Administration & overhead				.63			5.85				.53			7.20
Total							5.85							7.20

ing total MHPT at the coke oven (0.80) to the total MHPT which are needed at earlier stages to provide inputs to coke production—in this case coal handling. Since 0.57 tons of coke are required to produce one ton of hot metal (HM) and 0.84 tons of coal are required per ton of HM, 1.47 tons of coal are required to produce a ton of coke. Multiplying coal handling MHPT (0.05) by this input-output relationship gives the difference between total coke oven MHPT and cumulative MHPT through the coke oven stage.

- When alternative process steps are available, their share of total production is described in parentheses to the left of the direct MHPT requirement for that process step. For example, Table 5–2 shows that in 1980 16 percent of U.S. integrated output was made in open hearths and 84 percent in basic oxygen furnaces (BOFs), while Japanese integrated firms operated no open hearth furnaces.
- All labor involved in steel production, whether directly involved in operating a facility ("direct" input) or maintenance and material handling (the difference between "direct" and "total" for each process), from in-plant handling of raw materials to the shipment of the final product, is allocated to one or another of the processes.
- Labor requirements for overhead and head office staff (included in the total as "administrative and overhead") are calculated as a percentage share of the total, which is then applied uniformly to individual products.
- Manhours involved in unlisted plant processes are allocated to one of the processes listed on a user basis. For example, manhour requirements for the in-house production of oxygen are attributed to those processes that use the oxygen, such as the steelmaking furnace.
- Any manhours involved in producing intermediate products for sale are excluded.
- All labor employed in the production of the product concerned, including contract labor, is included.

These principles describe the method used in constructing the productivity tables and are the key to understanding them. Once the tables are constructed, it is possible to evaluate performance over time by comparing annual tables and to contrast international or intersectoral performance. Table 5–2, for instance, indicates a consistent Japanese advantage in the actual production of cold-rolled sheet, with a slight U.S. advantage in the "packaging and shipping" and "administrative and overhead" categories. In many cases, the Japanese advantage is slight. The most significant sources of the total Japanese MHPT advantage are in the hot end (with the extent of the advantage listed parenthetically): coke ovens (34%), blast furnaces (48%), raw steel production (31%), and slabbing mills (42%). For the latter two categories, the real locus of the Japanese advantage is described in the input-output relationships: the greater extent to which the Japanese use continuous casting (in 1980, 58% versus 20% in the United States) and the absence of open hearths in Japan.

Tables 5–3 and 5–4 provide similar information for the production of wire rods. Since this product is produced by both nonintegrated and integrated firms, MHPT requirements for wire rod can be compared between sectors as well as between countries. As one would expect, the mini-mill table (Table 5–4) is much simpler, since all processes up to the steel furnace are eliminated. The most striking insight provided by these tables is the extent to which mini-mills enjoy an advantage in both countries—even against Japanese integrated producers, generally recognized as the most efficient in the world. The mini-mill advantage actually exceeds total MHPT requirements through the raw steel stage for U.S. integrated producers, indicating that the superior performance of mini-mill producers cannot be attributed solely to the absence of the coke oven-blast furnace complex. The implications of the data presented in these tables will be discussed below; they are mentioned here only to illustrate the type of analysis that underlies the more aggregated treatment that follows.

PRODUCTIVITY TRENDS BY PRODUCT

Rather than reproduce a voluminous quantity of tables similar to those above, we have used this data base to develop overall descriptions of productivity trends, by product and by process. Here we will present and briefly discuss the results for several products: hot-rolled sheet, used extensively in the production of appliances and automobiles (15.5% of final U.S. shipments in 1980)[6]; cold-rolled sheet, which represents a further processing of hot-rolled sheet and constituted 16.5 percent of U.S. industry shipments in 1980; plate, 8.5 percent of 1980 shipments; hot-rolled bar, 12.6 percent; and wire rods, 3.4 percent. This list includes the most significant products in terms of their share of total domestic shipments. Hot-rolled bar and wire rods are commonly produced by both integrated and mini-mill producers; the performance of each of these sectors can therefore be compared in these product lines. Table 5–5 and Figure 5–1 compare historical trends in labor productivity (MHPT) for the production of several steel products in the United States and Japan. Table 5–5 also shows the rates of change of labor

Table 5–3. Comparative Manhours per Net Ton by Process: Wire Rod, Integrated 1980 (actual operating rates).

Japan										United States								
Process							Cumulative			Process							Cumulative	
Direct					Total	I/O	Direct	Total		Direct					Total	I/O	Direct	Total
					.04				Ore handling						.06			
		.02			.16				Sintering			.05			.25			
					.04				Coal handling						.05			
		.31			.53		.31	.60	Coke ovens			.48			.80		.48	.87
						1.55			Ore/HM							1.49		
						.76			Coal/HM							.84		
						1.27			Sinter/HM							.36		
						.52			Coke/HM							.57		
		.10			.34		.29	.93	Blast furnace			.27			.65		.56	1.33
						1.00			HM/CS							.80		
						.09			Scrap/CS							.36		
									Steel furnaces									
(1.00)	.14			.41					Melt:	(1.00)	.32			.57				
(0.00) -			-						OH	(0.16) .60			1.10					
(1.00) .14			.41						BOF	(0.84) .27			.47					
(0.42)	.21	.34		.25	.69		.63	1.62	I.C.	(0.80)	.11	.50		.31	.99		.95	2.05
(0.00) -			-						OH	(0.16) .18			.52					
(1.00) .21			.25						BOF	(0.84) .09			.26					
(0.58)	.20			.30					C.C.	(0.20)	.43			.87				
						1.060			CS/Billet							1.15		
(0.42)	.47	.20		.76	.32		.87	2.04	Billet	(0.80)	1.20	.96		2.00	1.60		2.05	3.96
						1.050			Billet/Rod							1.09		
		.78			1.10		1.69	3.24	Rod mill			.75			1.20		2.98	5.52
					.55			3.79	Package & shipping						.45			5.97
					.45			4.24	Administration & overhead						.48			6.45
								4.24	Total									6.45

Sources: See Appendix C.

Table 5–4. Comparative Manhours per Net Ton by Process: Wire Rod, Mini-mills 1980 (actual operating rates).

Japan Process Direct		Japan Process Total		Japan I/O	Japan Cumulative Direct	Japan Cumulative Total		United States Process Direct		United States Process Total		United States I/O	United States Cumulative Direct	United States Cumulative Total
Direct	*Direct sum*	*Total*	*Total sum*	*I/O*	*Direct*	*Total*		*Direct*	*Direct sum*	*Total*	*Total sum*	*I/O*	*Direct*	*Total*
							Electric furnaces							
(1.00) .45		.76					Melt:	(1.00) .41		.70				
(0.30) .45	.78	.48	1.16		.78	1.16	I.C.	(0.30) .28	.74	.35	1.12		.74	1.12
(0.70) .28		.36					C.C.	(0.70) .35		.45				
				1.035			CS/Billet					1.035		
(0.30) .62	.19	.91	.28		1.00	1.48	Billet	(0.30) .60	.18	.90	.27		.94	1.43
				1.050			Billet/Rod					1.075		
	.74		1.06		1.79	2.61	Rod mill		.75		1.25		1.76	2.79
			.41			3.02	Package & shipping				.40			3.19
			.45			3.47	Administration & overhead				.32			3.51
						3.47	Total							3.51

Sources: See Appendix C.

Table 5–5. Productivity by Product.

A. Absolute MHPT levels

	1958		1964		1972		1980	
	U.S.	Japan	U.S.	Japan	U.S.	Japan	U.S.	Japan
Integrated								
Hot-rolled sheet	9.16	27.03	7.85	14.40	6.22	6.17	5.36	4.42
Plate	10.89	28.64	9.73	15.70	8.33	6.89	7.73	5.26
Cold-rolled sheet	11.58	35.65	10.01	19.12	8.07	8.21	7.21	5.84
Hot-rolled bar	12.50	32.10	11.50	17.49	9.50	6.91	7.60	4.65
Wire rod	12.28	27.73	10.80	14.03	8.32	5.73	6.45	4.24
Mini-mills								
Hot-rolled bar	13.60	37.62	9.85	19.32	6.00	7.74	3.85	3.85
Wire rod	13.60	34.71	9.40	17.17	5.70	7.21	3.51	3.47

B. Annual rates of change in MHPT (%)

	1958–64		*1964–72*		*1972–80*		*1958–80*	
	U.S.	*Japan*	*U.S.*	*Japan*	*U.S.*	*Japan*	*U.S.*	*Japan*
Integrated								
Hot-rolled sheet	2.5	10.0	2.9	10.1	1.8	4.1	2.4	7.9
Plate	1.9	9.5	1.9	9.8	0.9	3.3	1.5	7.4
Cold-rolled sheet	2.4	9.9	2.7	10.0	1.4	4.2	2.1	6.5
Hot-rolled bar	1.4	9.6	2.4	10.9	2.7	4.9	2.2	8.4
Wire rod	2.1	10.7	3.2	10.6	3.1	3.7	2.9	8.2
Mini-mills								
Hot-rolled bar	5.2	10.5	6.0	10.8	5.4	8.4	5.6	9.8
Wire rod	6.0	11.1	6.1	10.3	5.9	8.8	5.9	9.9

productivity and includes more products than those illustrated in Figure 5–1.

Even for the production of different products, Figure 5–1 shows a striking consistency in the relationship between U.S. and Japanese performance. In every case the U.S. curves are essentially flat; this lack of progress is especially pronounced for the integrated sector. The Japanese performance, on the other hand, is generally described by an almost rectangular-hyperbolic curve; rapid improvements from inordinately high levels are followed by more gradual but nonetheless definite progress. In almost every case—mini-mill hot-rolled bar being the exception—1980 MHPT for Japan are lower than for the United States, indicating that the rapid improvement in performance by Japanese steel producers was not due merely to their original disadvantage relative to U.S. producers. The data and charts indicate a very pronounced U.S. advantage for every product in 1958, an advantage that had narrowed dramatically by 1964, disappeared by 1972, and become an absolute disadvantage by 1980.

Figure 5–1. Comparative MHPT by Product.

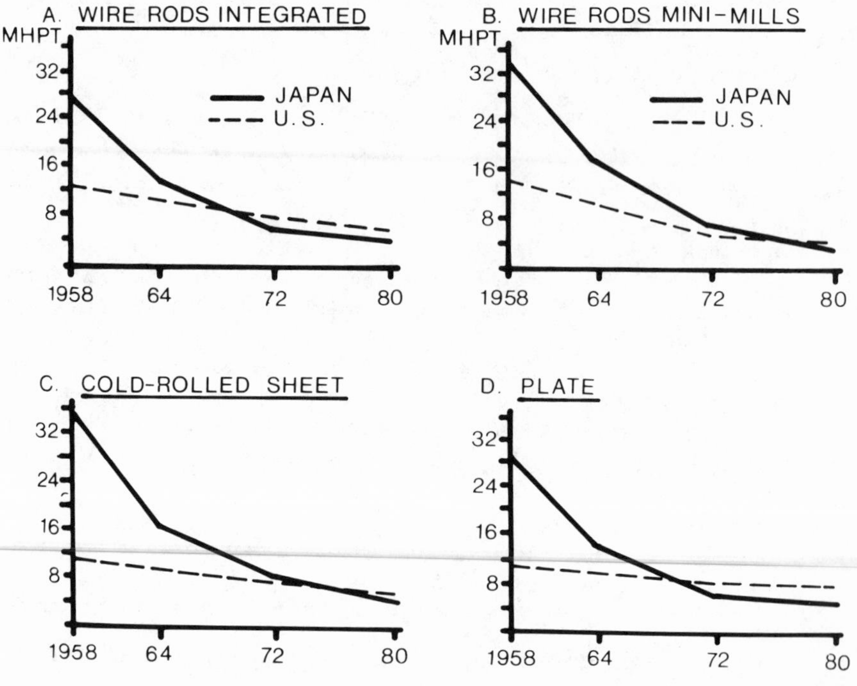

This overall trend can be broken down into its individual components, where the results are more varied. In certain product lines, U.S. integrated producers have maintained rough parity with their Japanese counterparts. This is the case for flat-rolled products, hot-rolled sheet (HRS) and cold-rolled sheet (CRS). Hot strip (sheet) mills are characterized by extensive economies of scale and were the target of significant investment expenditures in the early 1960s. U.S. hot strip mills thus tend to be newer than other integrated rolling facilities, and this helps to explain why U.S. integrated performance has to some extent kept pace with the Japanese in regard to flat-rolled products. The situation is almost exactly the reverse in regard to plate mills, where a dearth of investment on the part of U.S. producers has had predictable results in terms of performance. The record of U.S. integrated producers has been mixed in regard to mini-mill products, hot-rolled bar (HRB) and wire rod (WR). Several U.S. integrated producers made substantial investments in wire rod facilities in the early 1970s, and this is reflected in the fact that the highest rate of productivity growth for U.S. integrated producers is found with this product. Nonetheless, this was not adequate to give U.S. integrated producers an advantage in this product against any competitor; the U.S. mini-mill advantage over integrated in wire rod productivity is close to 50 percent.

The rate of productivity increase reflects the outstanding performance of Japanese integrated producers vis-a-vis their U.S. counterparts. As the data indicate, Japanese productivity increased about 10 percent per year from 1958 to 1972 and slowed to about 4 percent per year thereafter. Throughout the 1958–80 period Japanese productivity growth in integrated operations exceeded the U.S. rate in every product. It is noteworthy, however, that in the middle and late 1960s the rate of productivity growth in the United States rose as a consequence of increased investment and the implementation of new technologies, particularly the BOF. As was described in Chapter 3, however, this surge of investment did not lead to significant improvements in profitability and thus could not be sustained. By the 1970s, the rate of productivity growth had again slowed for integrated U.S. producers. The 1970s also evidenced a substantial decrease (over 50%) in the rate of Japanese productivity improvement in integrated operations—presumably a consequence of the depressed world steel market in the late 1970s and the termination of the prodigious expansion in Japanese steelmaking capacity. Even at the reduced rate, however, Japanese improvements in integrated steel productivity have continued to exceed those in the United States by a wide margin.

Perhaps the most striking feature of these product productivity data is what they tell us about mini-mill producers. Here the performance of the U.S. industry is quite good; Japanese electric furnace operations show more rapid productivity improvement, but this is largely due to the fact that they originally had much more room for improvement. American mini-mills have consistently achieved high rates of productivity growth, and by 1980 U.S. and Japanese mini-mill performances were essentially identical.

Comparison of the mini-mill and integrated sectors is even more striking. Two key points stand out. The first is that the nonintegrated sector in both countries is more efficient at producing the characteristic mini-mill products. The relative superiority of the mini-mill sector was evident as early as 1964 in the United States, while the reversal did not occur until the late 1970s in Japan. By 1980, however, mini-mills had gained a distinct advantage over integrated companies in Japan; in the United States, the mini-mill lead is overwhelming. The second point is in some ways even more significant: unlike their integrated counterparts, mini-mills have experienced no significant reductions in productivity improvements in the 1970s. During the 1970s the devastation of the world steel market, the end of major expansion, and the exhaustion of radical technological improvements have restrained productivity increases for integrated steel producers all over the world. The tempo of improvement in the mini-mill sector, however, has slowed hardly at all.

PRODUCTIVITY TRENDS BY PROCESS

The disaggregation of steelmaking into its component process steps makes it possible to describe the actual dimensions and structure of competitive efficiency. When combined with individual product data, information on productivity by process provides us with an even more useful microeconomic analysis of steel productivity.

The absolute manhour requirements for the outputs of various steelmaking processes are shown in Table 5–6, which presents data for the United States and Japan in 1958, 1964, 1972, and 1980. The rates of change in these variables are also presented in Table 5–6, while the results for several processes are described graphically in Figure 5–2. The overall pattern of productivity growth by process is similar to the pattern encountered in regard to specific products; since the production of any output merely represents the combination of several processes, this is not surprising. For U.S. integrated producers, improvements in productivity by process were strongest in the 1964–72 period. In the preceding period (1958–64) the overall rate of improvement had approached that found in 1964–72, while progress slowed drastically in the 1970s. In the case of Japan, the tempo of productivity improvement was quite

Table 5–6. Productivity by Selected Major Process.

A. Absolute MHPT levels

	1958		*1964*		*1972*		*1980*	
	U.S.	*Japan*	*U.S.*	*Japan*	*U.S.*	*Japan*	*U.S.*	*Japan*
Integrated								
Coke ovens	1.02	2.38	0.90	1.39	0.80	0.54	0.80	0.53
Blast furnaces	1.06	3.22	0.81	1.22	0.68	0.38	0.65	0.34
Steel making	2.20	5.12	1.82	2.20	1.12	0.78	0.96	0.66
BOF melt	1.17	1.73	0.86	0.90	0.54	0.46	0.47	0.41
CC	—	—	—	—	1.30	0.43	0.72	0.25
Slab mill	.80	2.26	0.76	1.05	0.66	0.75	0.63	0.68
Hot strip mill	1.22	4.18	1.15	1.99	0.98	1.22	0.93	0.79
Cold-sheet mills	2.22	6.27	1.93	4.31	1.70	1.87	1.57	1.33
Plate mill	3.20	5.84	3.14	3.70	3.11	2.05	3.07	1.59
Billet mill	2.75	3.27	2.55	1.85	2.25	0.82	2.00	0.76
Rod mill	2.90	5.85	2.60	3.15	1.75	1.55	1.20	1.10
Mini-mill								
EF	4.56	12.60	2.90	5.71	1.95	2.49	1.12	1.16
Melt	3.37	9.71	2.00	4.16	1.25	1.68	0.70	0.76
CC	—	—	—	—	0.80	0.69	0.45	0.36
Billet mill	2.80	4.36	2.20	2.34	1.30	1.10	0.90	0.91
Rod mill	2.90	6.36	2.30	3.70	1.65	1.82	1.25	1.06

B. Annual rates of change in MHPT (%)

	1958–64		*1964–72*		*1972–80*		*1958–80*	
	U.S.	*Japan*	*U.S.*	*Japan*	*U.S.*	*Japan*	*U.S.*	*Japan*
Integrated								
Coke ovens	2.1	8.6	1.5	11.1	—	—	1.1	6.6
Blast furnaces	4.4	15.0	2.2	13.6	0.6	1.4	2.2	9.8
Steel making	3.1	13.1	5.9	12.2	1.9	2.1	3.7	8.9
BOF melt	5.0	10.3	5.7	8.0	1.7	1.4	4.1	6.3
CC	—	—	—	—	7.1	6.6	7.1	6.6
Slab mill	1.0	12.0	1.7	4.1	0.6	1.2	1.1	5.3
Hot strip mill	1.0	11.7	2.0	5.9	0.7	5.3	1.2	7.3
Cold-sheet mills	2.4	6.1	1.6	10.0	1.0	4.2	1.6	6.8
Plate mill	0.3	7.3	0.1	6.1	0.2	3.1	0.2	5.7
Billet mill	1.3	9.1	1.6	9.7	1.5	0.9	1.4	6.6
Rod mill	1.8	9.8	4.8	8.5	4.6	4.2	3.9	7.3
Mini-mill								
EF	7.3	12.4	6.4	9.9	6.7	9.1	6.2	10.3
Melt	8.3	13.2	5.7	10.7	7.0	9.4	6.9	10.9
CC	—	—	—	—	6.9	7.2	6.9	7.2
Billet mill	3.9	10.0	6.4	9.0	4.5	2.3	5.0	6.9
Rod mill	3.8	8.7	4.1	8.5	4.4	6.5	3.8	7.8

Figure 5–2. Comparative MHPT by Process: Integrated Mills.

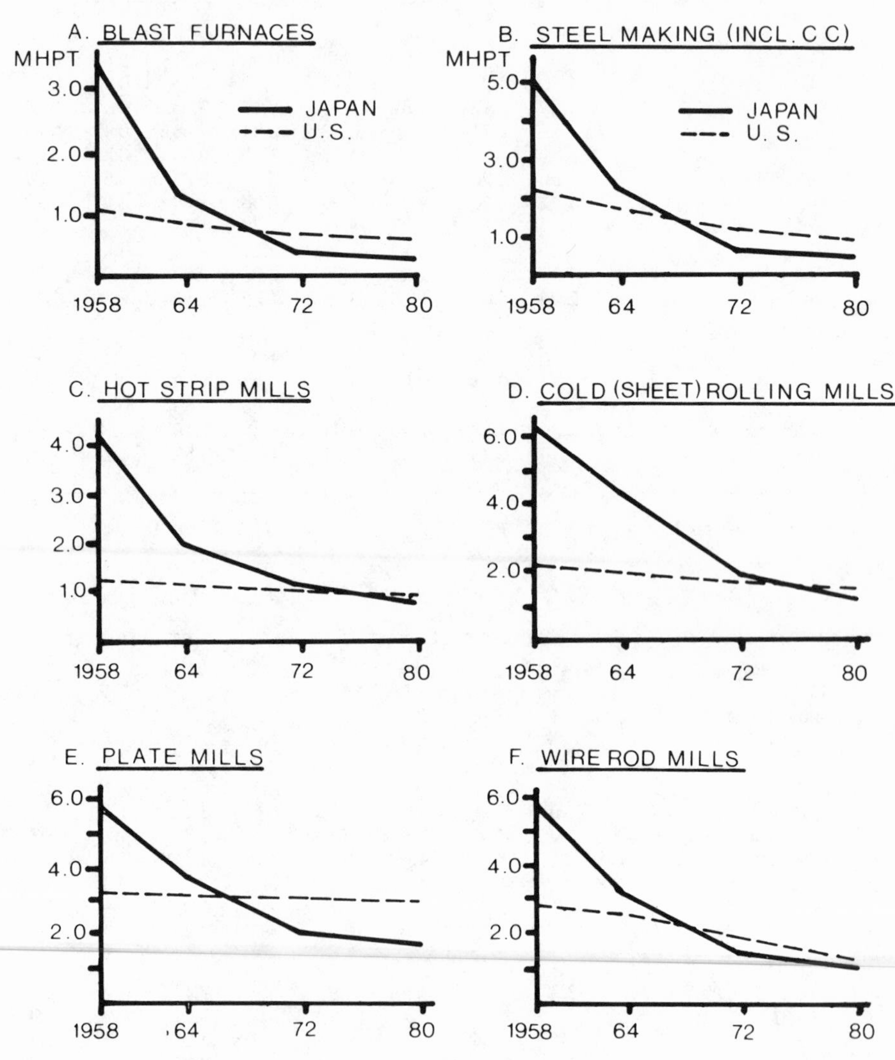

robust for integrated producers until the 1970s, and throughout the entire period the Japanese performance was far superior to that found in the United States. The overall parallel with product productivity results extends to the mini-mill sector as well. Here we find rapid growth for both the United States and Japan—not only from 1958 to 1972, but thereafter as well.

Integrated Producers

The general impression given by the process productivity data confirms the interpretation of the product data; detailed consideration of the results for each process leads to more interesting conclusions. If we turn first to integrated producers, we find that the Japanese performance was inferior to that of U.S. integrated producers for every process as late as 1964. In most cases this had been reversed by 1972. As early as 1972, the Japanese had a pronounced advantage in some processes (coke ovens, blast furnaces, steelmaking), while they faced a significant disadvantage in only one facility, the hot strip mill. By 1980 Japanese integrated producers had an absolute advantage, in some cases pronounced, in all processes but slab rolling. The U.S. advantage in this process is misleading, since slab mills had largely been replaced in Japan by the far more efficient continuous casting technology. The results suggest that prior to 1970 Japanese productivity efforts were concentrated on the primary end of the steelmaking process; after this date the Japanese emphasis shifted to finishing facilities. Whereas the overall slowdown from 1972 to 1980 is very pronounced in regard to primary processes, the average performance during this period for finishing mills is quite respectable. One consequence of this shift is that for the entire period from 1958 to 1980, the rate of productivity improvement is quite consistent among processes for Japanese integrated producers. Progress came on all fronts at a fairly uniform rate.

This consistency is lacking in the performance of U.S. integrated mills. In the early periods, U.S. productivity grew most rapidly in the primary end, as was the case in Japan. By the 1970s, however, U.S. productivity improvements were meager, except in the entirely new technology of continuous casting. This decade saw modest improvements in steelmaking performance, and substantial investment in wire rod facilities had a predictably positive effect. Otherwise, productivity growth ground to a halt. These trends describe an industry severely constrained in its ability to boost performance; this is the implication that must be drawn from the spottiness and fitfulness of the U.S. performance. Whether because of slow market growth, inadequate cash flow, or other reasons, U.S. efforts to improve productivity were much more selective and limited than in Japan.

In spite of far less spectacular rates of productivity growth, U.S. performance does not lag far behind the Japanese in the flat-rolled products, hot- and cold-rolled sheet and galvanized (not shown), that are the bread-and-butter items for the sales and profits of the major integrated companies. Major investments were made in hot strip mills in the early 1960s, and this is reflected in a slight but definite boost in the rate of productivity growth for the 1958–64 period. U.S. performance is also close to the Japanese in wire rod mills, which attracted substantial investment in the 1970s. In other finishing processes, however, U.S. integrated producers are now far behind; this applies to plate mills and bar mills (not shown).

In the primary or hot end—coke ovens through steelmaking—the Japanese advantage is overwhelming. In only one case, the production of raw steel in the BOF, is U.S. performance close to the Japanese level—a result due to the fact that both countries moved fairly rapidly, in roughly the same period, to exploit this new technology. In other primary processes, coke production and blast furnace operations, the United States now lags far behind Japanese standards. Since the 1950s, relatively little has been done to U.S. coke ovens in order to improve their performance—except shutting them down at an increasing rate. U.S. blast furnaces did show noticeable improvements up to the 1970s, but this was dwarfed by the prodigious advances made in Japan, which became the world leader in blast furnace technology before it claimed this status in any other process.

The respectable U.S. performance in the BOF has been offset by the failure to adequately employ continuous casting technology and by the anachronistic operation of some open hearths. In 1981, for instance, the U.S. open hearth share of production was 11 percent, while there were no open hearths operating in Japan. In the same year, over 60 percent of Japanese steel was continuously cast, while this was the case for only 22 percent of all U.S. steel production. Even this low figure exaggerates the prevalence of continuous casting in the integrated sector, since minimills, which use continuous casting to a far greater extent, are included in this calculation. Thus the respectable U.S. performance in terms of the BOF disappears when all primary processes through steelmaking are aggregated. For the combined processes of coke production, pig iron production (including sinter), and steelmaking (including continuous casting), 1980 Japanese manhour requirements averaged 64 percent of levels required in the United States.

Having compared Japanese and U.S. integrated performance in the finishing processes and the hot end, we need only discuss those intermediate processes that link these two major process categories, namely, primary rolling, in which ingots—the traditional output of the steelmaking stage—are reheated and rolled into semifinished shapes (blooms, slabs, billets) for further rolling into finished products. The application of continuous casting technology is presently eliminating this entire process; molten steel is being poured directly into semifinished shapes rather than into ingots which must be rerolled. Hence, the discussion of this "stage" is a bit confusing. As is shown in Table 5–2, in 1980 U.S. integrated producers had a slight advantage over the Japanese for one primary rolling facility— the slab mill. This was the only facility for which this was the case. However, for another primary rolling process, the billet mill, the U.S. disadvantage was relatively greater than for any other facility except continuous casters. Thus it is the link between the primary processes and the finishing processes which exhibits the most extreme and disparate results in terms of productivity. This confusing quantitative result is one consequence of the corrosive action of technical change, so that qualitative analysis is required to interpret the quantitative result.

The dominant factor that must be considered in the analysis of the intermediate process stage is the overwhelming Japanese lead in continuous casting. Since this technology replaces primary rolling per se, it reduces the relative significance of differences in primary rolling performance between the United States and Japan. In terms of continuous casting itself, the Japanese advantage in MHPT requirements is greater than for any other process, presumably reflecting differences in scale, operating practices, and experience using this technology. The U.S. advantage in slabbing is an anomaly; Japanese slabbing has been replaced by continuous casting for all but the more specialized products—that is, products that require more labor for inspection, scarfing, grinding, and so on. In the case of billet mills, U.S. practices generally involve ingot casting, the rolling of ingots into blooms, and the rolling of blooms into billets. Japanese ingot casting practices (e.g., "tailoring" ingot size to billet shape) often make it possible to skip the blooming stage. U.S. practices thus entail greater yield losses and higher expenditures of labor to rectify the defects associated with extensive rerolling.

While the effects of international differences in intermediate processes must be discounted to account for the impact of continuous casting, the impact of differing performance at the intermediate process stage can be quite significant for one product relative to another. All U.S. integrated products must cope with the relatively inefficient hot end. Once crude steel has been produced, however, relative performance in the intermediate and final processes can either boost or retard productivity growth by product. As can be seen in Table 5–5, the overall U.S. disadvantage in 1980 was greatest in regard to hot-rolled bar, where Japanese MHPT requirements were only 62 percent of the U.S. level. This difference is due to the fact that there has been little investment in this facility among U.S. integrated producers—the bar market has to a great extent been surrendered to mini-mills—and to the fact that bars are made from billets, so that poor bar mill performance is compounded by poor billet mill performance. In plate and wire rod, Japanese labor requirements were 68 percent of the MHPT necessary in U.S. integrated mills—an identical result that is reached via different routes. There has been substantial investment in wire rod mills, but this has been offset by reliance on billets; relatively good slab mill performance, on the other hand, has compensated for meager investment in plate mills. Finally, those products in which the U.S. disadvantage is smallest (hot- and cold-rolled sheet, where Japanese manhour requirements are 82 percent and 81 percent of the U.S. level) have been characterized by more adequate investment and the use of slabs as opposed to billets. Thus the constellation of processes required for the production of a specific product determines total manhour requirements and can effectively nullify the impact of major investment in the finishing process associated with that product. This is shown clearly in the case of wire rods.

The process disadvantages for U.S. integrated producers are structured in such a way that the absolute disadvantages are greatest for primary and intermediate processes. These disadvantages are then exacerbated if the U.S. finishing processes are highly unproductive in relation to Japanese performance (e.g., bars and plates). Alternatively, this disadvantage is unaffected if the American finishing processes are relatively less inefficient (e.g., hot- and cold-rolled sheet). In the former case, U.S. performance by product is highly deficient relative to the Japanese, while in the latter case the difference, by product, is reduced. Regardless of performance at the finishing end, however, Japanese integrated steelmakers enjoy a substantial advantage, applicable to every product, through the semifinished stage. This Japanese advantage is due primarily to raw materials processing and to much more extensive use of continuous casting.

While performance in the relevant series of process steps is clearly the principal determinant of productivity by product, product totals cannot be derived merely by adding up the subtotals for the component processes. This is due to the fact that operating performance may differ in terms of yields and the fact that intermediate outputs can in some cases be produced by several different techniques—for example, continuous casting vs. primary rolling. The whole, in a sense, is more complex than the sum of its parts; the parts must be weighted in terms of the extent to which they are used and the yields with which they are associated.

For integrated producers, this is illustrated in Table 5–7, which describes the sources of productivity improvements historically for a representative integrated product, cold-rolled sheet. This table refers to three time periods (1958–64, 1964–72, and 1972–80) and lists absolute changes in MHPT, by process, for this product. Input use refers to changes in the proportions of primary inputs used in the production of raw steel; thus a negative input use value implies that a higher proportion of the output of this process is being employed. Specifically, the input use entries in Table 5–7 reflect a decrease in the use of scrap, especially in Japan, due to the increased use of the BOF. Table 5–7 also shows the effects of changes in techniques: the replacement of the open hearth (more BOFs) and the replacement of ingot casting and slabbing (more continuous casting, CC). The impact of increases in yield is shown in a separate line after the continuous casting entry, since continuous casting is the most important single source of improved yields.

For the 1958–64 period, productivity improvements in the integrated U.S. industry principally occurred in four areas: blast furnaces, steelmaking, cold finishing, and overhead. Advances in the steelmaking stage, which were the most pronounced, stemmed from a combination of factors, especially the initial application of the BOF and increases in the scale and efficiency of open hearths. By contrast, Japanese productivity increases in this period were quite sizable and widespread. This outstanding performance, which admittedly refers to a relatively inefficient 1958 base, would have been even more striking except for the higher Japanese use of hot metal (molten iron) rather than scrap—as is reflected in the high negative input use value. Decreased reliance on scrap boosted labor requirements not only for blast furnaces but for coke ovens and sinter plants as well.

From 1964 to 1972, U.S. productivity improvements were greater and more evenly dispersed than in the preceding period. Nonetheless, they were still concentrated in the steelmaking area; both open hearths and BOFs were improved, while significant gains were realized via the switch from the open hearth to the more efficient BOF. This period witnessed the first introduction of continuous casting, with mixed results in the United States: advantages from the replacement of slab mills were exactly offset by disadvantages from the elimination of ingot casting. The slight advantage derived from continuous casting was associated with improved yields. In the case of Japan, productivity improvements were once again more substantial and more consistent. This was the period of the major expansion in Japanese capacity, when several large new mills were built. Overall Japanese performance was again weakened by increased use of hot metal, although to a lesser extent than in the previous period. As in the United States, this period witnessed the beginning

Table 5–7. Comparative Process Sources of Productivity Improvements: Integrated, Cold-rolled Sheet.

	1958–64 U.S.	*1958–64* Japan	*1964–72* U.S.	*1964–72* Japan	*1972–80* U.S.	*1972–80* Japan
A. *Absolute MHPT improvement by detailed processes*						
Primary processes						
Coke ovens	0.08	0.66	0.06	0.52	–	0.01
Blast furnaces	0.23	2.04	0.13	0.96	0.03	0.04
Sinter, etc.	0.02	1.19	0.03	0.82	–	0.10
Input use	–	−1.05	−0.09	−0.35	−0.01	−0.13
Steel making (Non-CC)						
Open hearth	0.36	1.87	0.16	0.03	0.04	–
BOF: Better performance	0.04	0.48	0.21	0.54	0.07	0.03
More BOFs	0.12	1.83	0.64	1.39	0.19	0.08
Continuous casting						
Better performance	–	–	–	–	0.08	0.08
More CC	–	–	–	0.17	–	0.28
Yields	0.04	0.60	0.01	0.45	0.11	0.23
Primary rolling	0.05	1.50	0.12	0.32	0.03	0.05
Hot rolling	0.08	2.45	0.19	0.84	0.06	0.46
Cold finishing	0.29	1.96	0.23	2.44	0.13	0.54
Package & overhead	0.26	3.00	0.25	2.78	0.14	0.60
Total	1.57	16.53	1.94	10.91	0.87	2.37
B. *Share of total MHPT improvement, by major processes (%)*						
Primary processes	21	23	11	21	3	6
Input use	–	−6	−5	−3	−1	−5
Steel making (Non-CC)	33	25	52	18	35	4
Continuous casting & yields	2	4	1	6	22	25
Primary & hot rolling	8	24	16	11	10	22
Cold finishing	18	12	12	22	15	23
Package & overhead	17	18	13	25	16	25
Total	100	100	100	100	100	100

of the transition to continuous casting. The advantages of this technology were much clearer in Japan, however. While the net advantage to U.S. users of continuous casting was 0.01 MHPT, Japanese continuous casters reduced labor requirements for the production of cold-rolled sheet by 0.62 MHPT in this period.

During the 1972–80 period, U.S. productivity in cold-rolled sheet exhibited the same anemic growth found in other products. Steelmaking was once again the principal process source of productivity improvements. Continuous casting emerged in this period as a significant source of productivity growth due to improvements in continuous casting practice, replacement of slabbing mills, and improved yields. In the Japanese case, the rate of productivity growth slowed even more drastically than in the United States, although it must be recalled that by 1972 U.S. and Japanese MHPT requirements for cold-rolled sheet were more or less identical. From this common base, Japanese improvements far outstripped those found in the United States. The Japanese disadvantage in terms of input use was drastically reduced, while the spectacular improvements in the hot end slowed noticeably. Significant gains continued to be made in the finishing end, although the increased use of continuous casting became the principal source of productivity improvements. While U.S. producers showed some gains from continuous casting, the Japanese performance was far superior. In the United States, the MHPT benefits of continuous casting were a meager 0.08, abstracting from the impact of increased yields; in Japan, this figure was 0.36—over four times as great. Furthermore, the impact of yield improvement was more than twice as great in Japan as in the United States.

The results of this analysis are summarized in Table 5–7, which shows the dispersion of the sources of productivity growth as a percentage of the total. This table immediately indicates the more consistent performance of the Japanese. The dominance of the steelmaking process in the U.S. case is evident in each of the periods considered, while the Japanese emphasis shifted from primary processes in the early years to intermediate and finishing processes later on. Other than this very general insight, there is no evidence in the Japanese data that any process or group of processes was dominant in any of the three periods, much less through all of them. Obviously, the U.S. approach to productivity improvement was narrow, while the Japanese approach was more comprehensive and complete.

The cumulative effects of these patterns can be summarized in terms of the 1980 Japanese advantage over U.S. integrated producers for two products, cold-rolled sheet and wire rods (see Table 5–8). The total Japanese advantage in cold-rolled sheet amounts to about 15 percent; it would be far greater were it not for the fact that the Japanese use a

Table 5–8. Sources of Japanese Productivity Advantages over the U.S.: Integrated 1980.

	Cold-rolled Sheet		*Wire Rod*	
	MHPT	*Share (%)*	*MHPT*	*Share (%)*
Primary processes				
Coke ovens	0.16	12	0.15	7
Blast furnaces	0.34	25	0.32	14
Sinter, Etc.	0.12	9	0.12	5
Input use	−0.45	−33	−0.44	−20
Steel making (Non-CC)				
Open hearth	—	—	—	—
BOF: Better performance	0.08	6	0.08	4
More BOFs	0.18	13	0.17	8
Continuous casting				
Better performance	0.23	17	0.24	11
More CC	0.20	15	0.45	20
Yields	0.37	27	0.29	13
Primary rolling	−0.04	−3	0.80	36
Hot rolling	0.15	11	0.10	5
Cold finishing	0.22	16	—	—
Package & overhead	−0.20	−15	−0.07	−3
Total	1.36	100	2.21	100

higher proportion of hot metal, relative to scrap, than do U.S. producers. Furthermore, the status of the Japanese industry as an exporter entails higher manhour requirements in packaging, shipping, and administration. The U.S. advantage in these areas is due less to U.S. performance than to the differing structural characteristics and resource base of the U.S. and Japanese economies. In terms of actual production, the Japanese superiority is stark. Japan's strengths lie in every process except the rolling of slabs from ingots—a disadvantage that is somewhat specious due to the reasons mentioned above. Almost 60 percent of the Japanese advantage in cold-rolled sheet can be attributed to continuous casting and higher yields. Although its principal current advantage stems from continuous casting, Japan has a significant lead in all processes other than slab rolling.

The Japanese advantage in wire rod amounts to roughly 33 percent—more than twice the figure for cold-rolled sheet. In this case (Table 5–8) the effects of continuous casting and yield differences, which account for about 45 percent of the Japanese advantage, are compounded by the Japanese superiority in billet mill performance, which accounts for 36 percent of the Japanese advantage. For the production of wire rods, U.S. integrated producers do not have an advantage in any process. While many processes are the same as in the production of cold-rolled sheet, slight differences in process MHPT requirements represent different input-output relationships. The cumulative effect of disadvantages in all the relevant processes for wire rod production is devastating, in spite of the fact that substantial investment in wire rod mills themselves have kept the disadvantage in regard to that step at the relatively low level of 0.1 hours—less than half the disadvantage associated with the cold finishing of hot-rolled sheet. Hence, significant modernization of wire rod mills per se has had little effect on the competitiveness of integrated U.S. wire rod producers vis-a-vis their Japanese counterparts. Unfortunately for American integrated producers, this gloomy picture is even more pronounced in regard to mini-mill performance in wire rod production.

Mini-mill Producers

The productivity record of U.S. mini-mills is one of the strongest indications that this sector should not be subsumed within steel industry aggregates. Whereas the image presented by the integrated sector is a gloomy composite of lagging productivity in almost every process, U.S. mini-mills have held their own even against their Japanese counterparts. This is shown in Tables 5–4 and 5–6 as well as in Figure 5–3 and Table 5–9. As is the case with integrated producers, U.S. mini-mills began with a substantial advantage over their Japanese counterparts, so that mini-mill productivity growth in Japan appears more impressive than is the case in the United States. This is somewhat misleading, however, since Japanese productivity growth has slowed as its absolute level approached that of its American counterpart. Although calculated on a lower base, U.S. performance was quite robust during the period in question, so that in 1980 Japanese and U.S. mini-mills were running neck and neck in regard to the main mini-mill products (e.g., wire rod and bars; see Table 5–5) and in terms of the major process categories of mini-mill techniques. (See parts A and B of Table 5–6.)

Figure 5–3. Comparative MHPT by Process: Mini-mills.

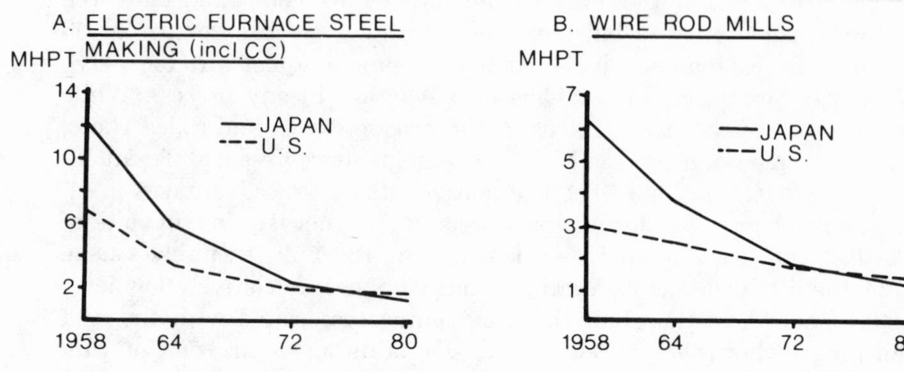

While the performance of U.S. and Japanese mini-mills are very closely matched, both are more efficient producers of the relevant products (bar and rod) than are their integrated competitors. The superiority of mini-mill techniques was established in the United States as early as 1964 (see Table 5–5), while this did not occur in Japan until the late 1970s. In the case of Japan, the eventual superiority of mini-mill techniques stemmed from the fact that productivity improvements in the integrated sector slowed drastically after 1974 but continued unabated in the mini-mill sector. The eventual result was that by 1980 the MHPT requirements of Japanese mini-mills were 18 percent less in wire rods than for their integrated counterparts, while the situation had been almost exactly the reverse in 1972. In the United States, the mini-mill advantage in wire rods amounted to 46 percent by 1980.

The process sources of mini-mill productivity improvements in Japan and the United States are shown in Table 5–9. It is noteworthy that while Japanese MHPT improvements were larger in the earlier years, beginning from a much less efficient level, this differential has since disappeared. Furthermore, the similarity in sources of productivity improvements, period to period, is striking. In both countries improved steelmaking has been the major source of productivity gains, although over time its importance has declined, while that of continuous casting and yields has increased. Improved primary and hot rolling have also made significant contributions to productivity gains in both countries. Mini-mill productivity improvement has been consistent across all major processes in both countries. For example, between 1972 and 1980, the

Table 5–9. Comparative Process Sources of Productivity Improvements: Mini-mills, Wire Rod.

	1958–64		*1964–72*		*1972–80*	
	U.S.	*Japan*	*U.S.*	*Japan*	*U.S.*	*Japan*
A. *Absolute MHPT improvement by detailed processes*						
Steelmaking (non-CC)	2.16	8.80	1.06	3.71	0.72	1.25
Continuous casting						
Better performance	—	—	0.17	—	0.24	0.19
More CC	0.54	—	0.52	0.83	0.22	0.47
Yields	0.21	0.47	0.14	0.36	0.08	0.18
Primary rolling	0.54	2.28	0.57	1.15	0.22	0.10
Hot rolling	0.55	2.66	0.70	1.88	0.40	0.76
Package & overhead	0.20	3.33	0.54	2.03	0.30	0.79
Total	4.20	17.54	3.70	9.96	2.18	3.74
B. *Share of total MHPT improvement by major processes (%)*						
Steelmaking (non-CC)	51	50	29	37	33	33
CC & yields	18	3	22	12	25	23
Primary & hot rolling	26	28	34	31	28	23
Package & overhead	5	19	15	20	14	21
Total	100	100	100	100	100	100

four major process sources each made roughly equal contributions to productivity in both Japan and the United States. As discussed above, such consistency across major process sources was characteristic of Japanese integrated performance but lacking in the U.S. integrated sector. U.S. integrated producers have attempted to remain competitive with mini-mill producers in wire rod production—an effort that has not been evident in regard to bar products. As was noted above, several large and modern wire rod mills were built by integrated companies in the early 1970s, yet the relative advantage of mini-mill techniques still increased between 1972 and 1980. This is strong evidence that the growth of the mini-mill sector is not due solely to failure on the part of integrated producers but is instead an objective, technology-based process of structural change—an interpretation also supported by the parallel performance of the mini-mill sector in both the United States and Japan during the 1970s. Technological changes in this decade have consistently favored the simpler processes of mini-mills; productivity improvements in electric furnaces outstripped those in integrated operations. In the United States, mini-mills use continuous casting far more extensively than do their integrated competitors.

Table 5–10. Sources of U.S. Mini-mill Advantages over Integrated: Wire Rod in 1980.

	MHPT	*Share (%)*
Primary processes and Input Use	1.27	43
Steelmaking (Non-CC)	−0.20	−7
Continuous casting:		
Better performance	0.23	8
More CC	0.63	22
Yields	0.19	6
Primary rolling	0.65	22
Hot rolling	−0.05	−1
Package & overhead	0.21	7
Total	2.93	100

The sources of the U.S. mini-mill advantage over integrated techniques in the production of wire rods (1980) are shown in Table 5–10. Not surprisingly, the absence of primary processes in mini-mill operations is the principal single source of their advantage; nonetheless, it is still less than the total advantage derived from performance differences in shared processes. In spite of the overall mini-mill advantage, integrated producers are almost as efficient in rod making itself (again the result of substantial integrated investment in this facility) and in steel production, where the electric furnace cannot match the scale efficiencies of the BOF. These slight integrated advantages, however, are eclipsed by the mini-mill advantage associated with continuous casting and, secondarily, by the fact that billet mills are more efficient in the nonintegrated sector.

CONCLUSION

The density of the statistical material presented above can tend to confuse rather than clarify; the reader who fails to see the forest for the trees is probably not alone. Having presented the raw data on productivity, at least in a summary fashion, we can now pause before proceeding to discuss the sources of productivity growth in Chapter 6. We are thus at a suitable point to recapitulate the most significant elements in the productivity record of the U.S. and Japanese steel industries.

Several of the conclusions drawn from these data confirm general impressions familiar to all observers of the world steel market. As expected, in the 1950s the U.S. steel industry had a significant advantage in labor productivity over its Japanese counterpart. This had been eliminated by the early 1970s, and the U.S. industry has continued to lose ground since then. Furthermore, within each national industry the mini-mill sector has become increasingly efficient compared to its integrated counterpart in the relevant bar and rod products. This may be somewhat surprising in the case of Japan, where the mini-mills did not gain the advantage until the late 1970s. In the United States, however, this occurred in the mid-1960s—that is, soon after the rapid growth in the mini-mill sector began. By 1980, U.S. mini-mills were far more efficient producers of commodity-grade bars and rods than were their integrated competitors.

The temporal pattern of productivity increase also corroborates expectations developed from an analysis of patterns in investment, market growth, and so on. Japanese productivity improvements in the integrated sector were quite robust through 1972 and respectable, albeit not spectacular, thereafter. For U.S. integrated companies, however, productivity improvements never approached the Japanese rate; somewhat more rapid progress was made between 1964 and 1972, when industry investment was at its highest postwar level, but this effort seems to have been completely exhausted thereafter. In both countries historical rates of productivity growth fell victim to the post-1973 stagnation in the world economy, which has been particularly evident in the steel market. Strikingly, however, productivity growth in the mini-mill sector slowed only slightly in the 1970s – clear evidence of the increasing technological attractiveness and aggressiveness of this sector. Any solace that U.S. integrated producers could draw from the tapering off of Japanese productivity growth has been offset by the increased efficiency of their mini-mill competitors.

In terms of individual products, U.S. integrated firms are most competitive in the product lines that have traditionally dominated steel markets in the automobile age: hot- and cold-rolled sheet. In 1972, in fact, these were the only products in which U.S. integrated producers outperformed (cold-rolled sheet) or more or less matched (hot-rolled sheet) their Japanese rivals. In the case of mini-mill products (bar and rod), U.S. and Japanese electric furnace operations had almost identical productivity in 1980. Both outperformed their integrated counterparts, and both showed consistent progress in labor productivity from period to period.

Differing performance in regard to specific products can be related to the constellation of processes that are employed to make that product. The spectacular gains made by Japanese integrated steel producers in productivity by product were matched in terms of processes. Up to the early 1970s, Japanese process improvements were dominated by outstanding progress in the hot end. Hence, the U.S. disadvantage is normally widest through the steel melting and casting stage (except for billets). Nonetheless, the Japanese made consistent progress in all processes. In the United States, the overall slow rate of productivity improvement, by product, is reflected in a sketchy and inconsistent record for individual processes. Faced with limited funds for investment, U.S. firms were forced to target investment at selected products and processes: steel furnaces (BOFs) and hot strip mills in the 1960s, wire rod mills and (to some extent) continuous casters in the 1970s. Unfortunately for U.S. firms, the consequences of such an approach have been mixed. It appears that extensive investment and significant productivity improvements in selected processes have often had only marginal effects on product productivity. This is seen most clearly in the production of wire rods with integrated techniques. U.S. integrated producers made substantial efforts to retain a large share of this market; through the 1970s this process experienced the greatest growth in productivity except for the new technology of continuous casting. Yet this effort had little effect on competitiveness. Weighted down by the burden of low productivity in related processes, especially billet mills, integrated producers of wire rods have fought a losing battle to remain competitive in this product.

This is a sobering fact. There is little likelihood that the funds available for steel industry investment will be much greater through the 1980s than has been the case in the 1970s. Yet, significant gains in productivity will be very difficult for integrated producers to achieve without substantial improvements in most if not all of the major processes required for steel production. A repetition of the wire rod experience will be unacceptable, but lack of funds would seem to make it unavoidable unless there is a major readjustment of investment priorities. This is a dilemma that will prove very difficult for integrated producers in the United States. Indeed, it is hard to see how it can be resolved without major structural changes. The one area where the data offer the hope of significant improvements for U.S. integrated producers is in regard to continuous casting. This appears to be true regardless of whether the potential of this technology is judged by the performance of mini-mills (U.S. and Japanese) or Japanese integrated producers. For example, between 1972 and 1980, the total benefit to Japanese integrated producers from more extensive use of continuous casting was 0.59 MHPT; the corresponding gain in the United States was 0.16 MHPT. These data present a paradox: the potential benefits of continuous casting seem obvious and are generally assumed to be so, yet they are not so evident in the documented experience of U.S. integrated producers. This paradox may in fact explain the perplexing slowness of U.S. integrated firms to implement this technology. Perhaps the benefits they have been able to achieve have not been as great as advertised; possible reasons for this will be discussed in the next chapter. Nonetheless, the argument could be made that the overall failure of U.S. integrated producers to keep pace with their Japanese competitors in the 1970s stemmed from the slow transition to continuous casting. Whereas the transition to the BOF had been aggressively pursued in the 1960s, this effort was not sustainable with the new technology of the 1970s. The effects on productivity growth in the integrated sector have been grim.

Notes

1. Peter F. Drucker, "Toward the Next Economics," in Donald Bell and Irving Kristol, eds., *Crisis in Economic Theory* (New York: Basic Books, 1981), p. 12.
2. There is an extensive literature concerning productivity measurement and analysis. A relatively recent collection is presented in John Kendrick and Beatrice Vaccara, eds., *Conference on New Developments in Productivity Measurement and Analysis* (Chicago: University of Chicago Press, 1980).
3. See, for example, Japanese Ministry of Labor (JML), *Labor Productivity Statistics Survey*, Annual 1958–80 (Tokyo: JML, 1960–82).
4. Harvey Leibenstein, "Allocative Efficiency versus 'X-Efficiency'," *American Economic Review*, 56 (June 1966): 392–415.
5. Throughout this book, the following abbreviations are commonly used, particularly in the tables and charts that deal with productivity, such as Tables 5–2 to 5–4: I/O (input-output), HM (hot metal or molten iron), CS (crude or raw steel), OH (open hearth), BOF (basic oxygen furnace), EF (electric furnace), IC (ingot cast-

ing), CC (continuous casting), HR (hot rolling), and CR (cold rolling).

6. Hot-rolled sheets (coils) are used in steel plants to produce virtually all flat-rolled products other than thick plates. In 1981 over 48 percent of total U.S. steel shipments were, at an earlier stage of production, hot-rolled sheets.

6 THE SOURCES OF PRODUCTIVITY GROWTH

... [T]he second great question in political economy: on what the degree of productiveness ... depends.[1]

– John Stuart Mill

The productivity record described in Chapter 5 is the product of myriad factors, from direct causes such as the installation of new equipment to the cultural milieu that defines the relationship between labor and management. An understanding of how such factors affect productivity performance will clarify the prospects for reversing the competitive decline of the American steel industry—the theme to which we now turn. Our principal intent in this chapter is to link the documented record of productivity growth to several of the factors that affected it.

The sources of productivity growth are highly interdependent, and this makes it difficult to uniquely relate past productivity improvements to specific causes. Even if this were possible, however, it would not provide an unambiguous guide to the best means of achieving improved performance in the future. The rate of productivity growth in the Japanese steel industry has clearly been superior to that achieved in the United States. Yet this does not necessarily imply that the American industry should seek to emulate the Japanese strategy of anticipating market growth and building massive tidewater plants to gain the maximum economies of scale. Raising productivity must be a central element in a firm's strategic orientation, but strategy must be appropriate to the business environment in which the firm functions. Without the Japanese environment, particularly rapid market growth, attempts to pursue the Japanese strategy are likely to produce disastrous results.

Another reason that knowing the past causes of productivity performance should not be confused with knowing the optimal strategy for improving future performance, is the fact that firms compete directly in terms of prices and costs rather than physical efficiencies. Measures that boost labor productivity may have negative effects on the cost of other inputs, particularly capital. Only when overall costs have been consid-

Table 6–1. Market Growth, Investment, and Productivity. (%, except where noted)

	1958–64		*1964–72*		*1972–80*	
	U.S.	*Japan*	*U.S.*	*Japan*	*U.S.*	*Japan*
A. *Market growth*[a]						
Domestic Consumption	+ 3.4	+14.3	+ 2.8	+13.4	+ 0.8	+ 0.5
Shipments	+ 1.7	+17.6	+ 2.1	+14.6	− 0.4	+ 1.4
B. *Investment rates*[b]						
Investment/ton shipped (1980 $)	43.4	81.3	46.9	56.4	39.6	47.6
Net capital stock growth	+ 2.1	+29.2	+ 2.5	+12.2	+ 0.2	+ 6.2
Investment/net capital stock	10.2	26.6	11.1	20.1	8.9	13.9
C. *Productivity growth*[c]						
CRS	+ 2.4	+ 9.9	+ 2.7	+10.0	+ 1.4	+ 4.2

a. Annual percentage rate calculated from 5-year average around end-point of period for all steel products.

b. Annual average for each period in 1980 dollars. Includes environmental and nonsteel expenditures. Japanese data converted using average annual exchange rates.

c. Annual percentage rate.

Sources: American Iron and Steel Institute (AISI), *Annual Statistical Reports* (New York and Washington, D.C.: AISI, various years); Japan Iron and Steel Federation (JISF), *Statistical Yearkbooks* (Tokyo: JISF, various years); *Monthly Reports of Iron and Steel Statistics* (Tokyo: JISF, various months); Japanese government statistics; and D. F. Barnett, *Economic Papers on the Steel Industry,* (Washington, D.C.: AISI, 1981).

ered is it possible to define the appropriate strategy for improving performance. Productivity is only one component of the total picture, though arguably the most important.

This chapter will begin with the most crucial element in the business environment faced by an industry—the trend in demand. This theme will be followed by an analysis of investment, after which we will turn to several of the immediate determinants of productivity performance: economic choices (e.g., concerning inputs), technology, and scale. The lessons that arise from this analysis will be applied in Chapter 7, which discusses the prospects for revitalizing the performance of the U.S. steel industry in the broader contexts of overall cost competitiveness and structural change.

MARKET GROWTH

Of all the conditions that define the environment in which a firm operates, none is more important than the trend in demand. Rapid market growth is a powerful elixir for many aspects of corporate performance; as far as productivity is concerned, it has dramatic, though indirect, consequences on several fronts. Booming markets foster the construction of new plants, which are generally larger and more advanced technologically; both of these characteristics reduce labor requirements. Besides these effects on the physical structure of a firm's facilities, rapid market growth also tends to boost profits and cash flow. The availability of funds for investment then accelerates the modernization of existing facilities. Finally, rapid market growth makes it easier to maintain high operating rates, lowering unit costs and thus improving profitability. All of these characteristics contribute to productivity growth.

Table 6–1 restates the evidence on market growth presented in Chapter 2, linking this to trends in investment and productivity. Since the late 1950s, demand has grown at a robust pace in Japan (about 10% annually) and at an anemic rate in the United States (about 2% per year). This difference implies contrasting potentials for productivity improvement.

Through much of the postwar period, rapid growth in Japanese steel consumption allowed the Japanese industry to sustain spectacular rates of investment. This was particularly true in the 1958–64 period, when the Japanese industry began to emerge as a major force in the world steel market. Booming markets attracted huge amounts of capital, with beneficial results for productivity. In the case of the United States, a stagnating market has been associated with a consistently weak pattern of investment, with little variation from period to period. Similarly, there is little variation in the overall trend in productivity.

The booming Japanese market collapsed in the 1972–80 period. In fact, as Table 6–1 shows, consumption actually grew more rapidly in the United States than in Japan during those years. Japanese productivity growth also fell drastically, although much less than consumption or shipments. This is an anomaly; in every other case the link between market growth and productivity improvement appears to be very strong. The respectable Japanese record of investment and productivity performance during the 1972–80 period suggests that it may be possible to gain significant productivity improvements even in a slow-growth environment. The Japanese industry was able to achieve this result by seeking exports to compensate for domestic demand, maintaining a high rate of investment, and exploiting the momentum generated by spectacular productivity growth in the preceding periods.

Nonetheless, a more pessimistic interpretation of the Japanese performance during the last decade is even more plausible. The slowing of Japanese productivity growth from the 1960s to the 1970s could be viewed as an indication that the Japanese industry has not been exempt from the effects of weak market growth. While the momentum of high-growth markets has sustained respectable rates of investment and productivity improvement over the past decade, there is as yet little indication that the strategy of the Japanese steel industry has shifted to a slow-growth orientation. Excess capacity, the bane of any industry faced with a flat trend in demand, is now a crushing burden for Japanese steel producers. Japanese operating rates have averaged well below 70 percent since 1975, while exports have averaged about 35 percent of total production during the same period.[2]

Although the U.S. steel industry has faced the problem of slow market growth for most of the postwar period, it has also failed to adopt a slow-growth strategy. Weak markets, low levels of investment, and poor productivity improvement have represented a vicious cycle for U.S. steel producers. Furthermore, the negative consequences of relatively stagnant markets have been exacerbated by the structure of the U.S. industry. Given the relatively large number of major U.S. producers, slow market growth has offered limited opportunities to the individual firm for adding capacity and increasing scale. By the late 1950s, the large size of the U.S. market had ceased to have a progressive effect on the scale of steel operations; once adequate (or even more than adequate) capacity was in place, slow market growth locked the existing scale and locational structure of the industry into place. One consequence of this was an increase in the marginal capital cost of replacing and modernizing existing capacity. Replacement is facilitated by the addition of new capacity, since the capacity of a new plant can usually be boosted to permit the retirement of outmoded facilities at little increase in cost above the level necessary to add capacity for market growth alone. With smaller total investment and more costly replacement, economies of scale are lessened, as are the prospects for significant technical changes. This in turn retards growth in labor productivity.

Slow market growth has made it more difficult for the U.S. steel industry to make consistent gains in labor productivity, and similar conditions may constrain the future performance of the Japanese industry as well. Throughout the postwar period, the U.S. industry has been haunted by the close correlation between productivity and market growth described in Table 6–1. Since there are few prospects of significantly boosting the secular trend in domestic steel demand, the efforts of American companies must be directed at loosening the link between productivity and market growth. In a sense, this chapter and the next represent a commentary on the appropriateness of the industry's efforts to boost performance, *given a slow-growth market.*

INVESTMENT

Aggregate data on the level of investment were provided in Chapter 3, and similar data are presented in Table 6–1. While these data suggest the superior performance of the Japanese industry, they must be disaggregated to see the implications of historical spending patterns for productivity growth. Different levels of capital expenditure are associated with different types of investment and thus with different rates of productivity improvement. Investment expenditures will not generate rapid productivity improvement if they are devoted to maintenance projects such as relining blast furnaces or to mandated, nonproductive expenditures such as pollution controls. Environmental expenditures have absorbed approximately 20 percent of investment in the U.S. steel industry since 1976.[3] Maintenance and regulatory expenditures usually cannot be postponed, even though they may not boost performance. Significant productivity improvements are associated only with discretionary investments—those that improve technology or increase scale, whether incrementally or via major projects.

A rough approximation of the extent to which investment is devoted to discretionary expenditures can be discerned in the relationship between the level of investment and the net capital stock. Insofar as the real value of the net capital stock is increasing, funds are being invested above the level needed to maintain facilities. Positive net investment is thus one indication that productivity-enhancing expenditures are being made. Alternatively, a stagnant or declining value for the net capital stock is an indication that investment funds are being consumed by projects that are unlikely to generate overall productivity improvements, regardless of their specific benefits.

Table 6–1 presents evidence on the historical relationship between investment, net capital stock (both in constant 1980 dollars), and productivity in the U.S. and Japanese steel industries. For both countries, capital stock estimates were derived from investment data using a twenty-year, straight line depreciation schedule. We thus assume that investments made more than twenty years previous to the year in question do not contribute to the net capital stock, even if the facilities that embody the investment are still in operation. In other words, the assumption is made that the economic life of steel facilities is, or should be, twenty years, regardless of their physical life. On that basis, the 1980 value of the net capital stock in the U.S. steel industry was approximately $38.6 billion; in Japan, it was $70.4 billion. Since Japanese steelmaking capacity only slightly exceeded U.S. capacity in 1980, the tremendous difference in net capital stock is a rough indication that U.S. facilities are almost twice as old as Japanese equipment.

Table 6–1 clearly illustrates the consistently weak pattern of U.S. investment and performance relative to the Japanese. The salutary effects of the boom in the Japanese steel industry prior to the mid-1970s are obvious. Its net capital stock grew at an incredible pace through 1972, fed by a very high ratio of investment to net capital stock. Productivity growth during this period was also striking. In the U.S. case, modest increases in the net capital stock supported modest increases in productivity from 1958 to 1972. By the 1970s, U.S. investment had fallen to a level that barely maintained the value of the net capital stock.[4] In fact, these estimates indicate that in real terms the U.S. steel industry has been disinvesting since the value of the net capital stock peaked in 1970; investment levels have not been adequate to maintain, much less significantly modernize, existing facilities. Theoretically, the funds available for discretionary investment are now negative; such projects are carried out only at the expense of necessary maintenance expenditures. This trend of disinvestment is actually understated in the data on U.S. investment, since these figures include both governmentally mandated expenditures, which do not increase the productive capital stock, and nonsteel investments, which have been increasing in recent years and which now constitute roughly 30 percent of total investment by steel firms.[5]

Real net investment has also declined in the Japanese steel industry; the ratio of Japanese investment to the net capital stock during the 1972–80 period was only slightly above the level achieved in the United States from 1958 to 1972. While the capital stock grew at a more rapid rate than was the case in the United States, most of this growth occurred in the first half of this period. By 1978, the net capital stock had also begun to decline in Japan—roughly ten years later than in the United States. This is another indication that the conditions faced by the Japanese steel industry had changed drastically by the late 1970s.

Unfortunately, there has been no change, except perhaps for the worse, in the environment faced by the U.S. industry. The declining value of its net capital stock indicates that the industry has been investing at a level inadequate to maintain the economic viability of its facilities, assuming a twenty-year economic life. If the U.S. industry has in fact been in a maintenance mode, this would partially explain the industry's deteriorating productivity performance. It implies that the industry has been standing still in terms of the rate at which its facilities are replaced and thus the rate at which new technologies are implemented. Obviously, the pattern of investment has been uneven; some facilities have been replaced at a more rapid rate. Yet an advance on one front has been more than offset by retreat on others.

This conclusion is buttressed by an analysis of historical trends in the capital-labor ratio, which can be used to describe the nature of investment. Chapters 2 and 3 described the different implications of capital-widening and capital-deepening investment. The former will affect productivity only slightly, even though it may boost the value of the net capital stock. Capital-deepening investments, on the other hand, may have a profound impact, since they represent the replacement of existing

Figure 6–1. Labor Productivity and the Capital Stock, 1950–1981.

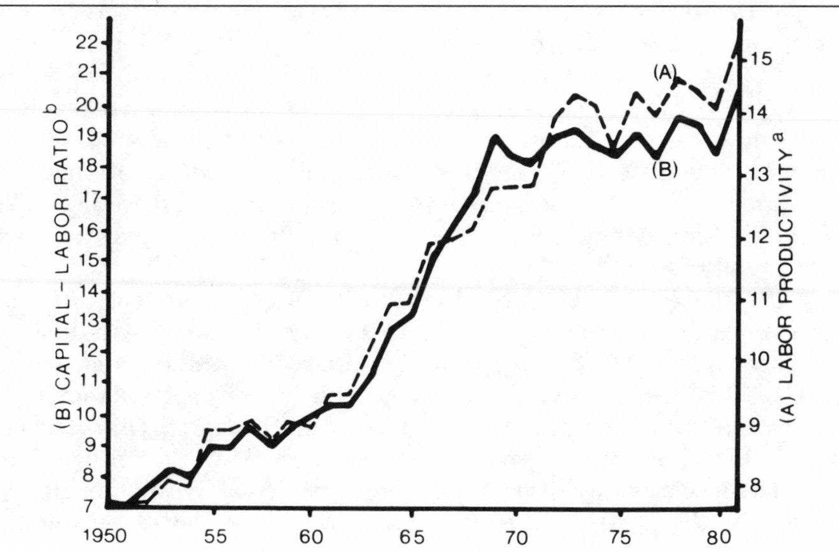

a. Calculated in terms of tons of raw steel produced per 100 hours of labor time.
b. Calculated in constant 1972 dollars per labor hour weighted by capacity utilization and excluding nonsteel and nonproductive investments from the measured capital stock.
Sources: Both productivity and capital stock calculated from financial data presented in American Iron and Steel Institute (AISI), *Annual Statistical Report* (New York and Washington, D.C.: AISI, various years). Operating rate from same source and from Council on Wage and Price Stability (COWPS), *Prices and Costs in the United States Steel Industry* (Washington, D.C.: U.S. Government Printing Office, 1977).

technology by new. Increases in capital-labor ratios are often used as a proxy for capital-deepening investment; Figure 6–1 relates postwar labor productivity in the steel industry to the industry's utilized capital-labor ratio. It should be noted that this illustration measures aggregate labor productivity (in terms of raw steel produced) instead of labor requirements for a single product, as in Table 6–1. Furthermore, Figure 6–1 adjusts the capital-labor ratio to take account of operating rates. Such an adjustment is necessary because the cyclical character of steel production causes significant fluctuations in labor productivity. One way to net out business cycle effects is to correct the capital stock value for utilization, as is done here. It should be noted that without adjustment for capacity utilization the capital-labor ratio is only weakly related to labor productivity.

Once this adjustment is made, however, Figure 6–1 shows clearly that labor productivity has been closely related to the utilized capital-labor ratio. Moreover, this graph can be interpreted to describe the link between the pace of productivity improvement and the nature of investment, that is, whether investment is devoted to capital-widening or capital-deepening projects. Much of the U.S. investment in the 1950s—at a time when cash flow was adequate and the industry was not threatened by imports—was unfortunately of the capital-widening type, increasing open hearth capacity by almost 50 percent while having little effect on the technological character of this process (although scale was increased). In general, breakthrough technologies were not sought, and productivity growth was correspondingly limited.

The rapid rise in the capital-labor ratio during the 1960s, reflecting capital-deepening investment, is associated with the industry's transition to the basic oxygen furnace (BOF). Investment reached its peak in this decade, and the rapid growth in the capital-labor ratio was matched by the pace of productivity improvement. Unfortunately, the U.S. industry did not gain any competitive advantage from the high levels of investment and productivity achieved during the 1960s. By the 1970s, the surge of investment achieved during the preceding decade could not be maintained. Investment in the U.S. steel industry became largely maintenance oriented; where major investment projects were undertaken, the results were often disappointing—e.g., new rod mills. For the U.S. steel industry, as for the economy as a whole, productivity improved much more slowly in the 1970s than in previous decades. In the case of steel, this was connected to the fact that the capital-labor ratio was basically stagnant throughout this decade; that is, there was very little capital-deepening investment. Hence, the industry's poor record of productivity growth during the past decade is linked to its low level of investment and to the related fact that investment has not been of the capital-deepening type.

It may be the case that the dearth of funds has made it impossible for firms to undertake discretionary, capital-deepening investment projects. Yet there is little evidence that the U.S. steel industry has seriously confronted the dynamic implications of this pattern. In particular, the industry has chosen to hold on to capacity that it could not afford to maintain adequately. This has caused limited investment funds to be dispersed among a large number of slowly declining plants. Reductions in capacity would have enabled limited funds to be targeted toward the most viable facilities–although even this strategy would have only partially offset the negative consequences of stagnating or declining net capital stock. Instead, the U.S. steel industry offers many examples of relatively modern facilities located in plants that are dominated by outmoded operations.

The strategy of widely dispersing investment stems from the industry's engrained commitment to maintaining capacity for the boom rather than for normal levels of demand. As a result, the potential benefits of modern equipment are dissipated by the burden of older facilities. Furthermore, the decision to keep inefficient plants operating boosts the share of investment that must be devoted to maintenance or environmental retrofits and thus limits the sums available for major projects. Finally, the maintenance of excess capacity lowers average operating rates. Low operating rates directly reduce both productivity and profitability; low profits then restrict the funds available for productivity-enhancing investments.

THE IMMEDIATE DETERMINANTS OF PRODUCTIVITY GROWTH

Investment expenditures generate productivity growth only indirectly. In order to affect performance, capital expenditures must be embodied in new equipment. This brings us to the actual production process. There are three broad categories of factors that directly determine productivity performance: technology, scale, and what we will call "economic choices." In what follows, each of these factors will be discussed in turn and related to the record of productivity performance presented in Chapter 5. Their relative importance will be evaluated in Chapter 7.

Economic Choices

The term "economic choices" refers to differences in operating practices that are strongly influenced or determined by relative prices (especially input prices) or by structural characteristics of the market in which an industry operates. Different market conditions will foster different economic choices—a point that in a sense confuses the analysis of comparative productivity performance. Operating practices that represent a relative disadvantage for an industry in terms of productivity may be rational procedures given conditions in either factor or product markets. There is little point in arguing that productivity could be increased by altering economic choices in such cases, since improved productivity would entail higher costs or reduced sales. This interpretation applies to the three areas in which economic choices provide the American steel industry with an advantage vis-a-vis its Japanese rival: input choices, packaging and shipping, and overhead (administrative and clerical).

In regard to the choice of inputs, the extent to which an industry uses scrap is the key to productivity performance. Total reliance on scrap enables a firm to completely avoid the processing steps of agglomeration, coke production, and ironmaking. This economic choice provides mini-mills with one of their chief productivity advantages over their integrated rivals. Even within the integrated universe, however, greater use of scrap can have significant effects on performance and investment requirements. Faced with a bottleneck in coke production, for instance, a firm may choose to invest in new coke oven batteries. Alternatively, greater reliance on scrap (implying a reduction in the hot metal ratio) may alleviate the bottleneck, postponing or eliminating the need for coke oven investment. Since scrap supplies are relatively abundant in the United States, the price of scrap tends to be relatively low, encouraging greater reliance on scrap. Conversely, relative scarcity in Japan generates a different price structure, encouraging greater reliance on hot metal. This represents a productivity advantage for the United States that does not imply inefficiency on the part of Japanese steel producers.

Iron ore is also relatively abundant in the United States. Paradoxically, Japanese steel producers have lower iron ore costs, since they obtain their iron ore from countries that have a comparative advantage in iron ore vis-a-vis the United States. Nevertheless, economic choices concerning iron ore use appear to represent productivity disadvantages for the Japanese.

Both U.S. and Japanese integrated producers now rely almost completely on some form of preprocessed iron ore input. But while both countries have moved away from unprocessed ores, there is an enormous disparity in the choice of upgrading technique. In 1980, sinter constituted 82 percent of the burden in Japanese blast furnaces, while the sinter share was only 26 percent in the United States. By the same token, pellets made up 67 percent of the furnace feed in the United States and only 16 percent in Japan. The U.S. preference for pelletizing is an economic choice; U.S. firms own large reserves of low-quality ores that are upgraded via pelletizing, a mine site operation. The Japanese, on the other hand, generally do not own iron mines. Since their industry had been expanding, sintering lines could be designed into newly constructed steel mills at a relatively low cost. Labor requirements for sintering plants at the steel mill are counted in steel industry labor hours, while mine site pelletizing is not. This discrepancy increases manhours per ton (MHPT) in the Japanese industry, although it does not represent an inefficiency in the broader sense.

In addition to the effects of input choices, the Japanese steel industry also appears to incur a productivity disadvantage in final material handling (packaging and shipping) and overhead. This is connected with the fact that the Japanese industry is highly dependent on exports. An export orientation complicates the sales effort and increases the amount of labor that must be expended to prepare output for international shipment and more extensive handling. As a result, the decision of Japanese producers to seek significant export sales—an economic choice—contributes to higher labor requirements at these stages of the production process.

In general, the factors that we have described as economic choices have a rather peculiar status. While economic choices have significant consequences for productivity, their effects are restricted to specific entries in the productivity tables presented in Chapter 5. Technology and scale, on the other hand, have pervasive effects, so much so that their impact cannot be immediately discerned. In a sense, economic choices reflect the inherent comparative advantages of an industry. Comparative advantage (i.e., the structure of resource endowments) determines the pattern of relative prices, which in turn determines decisions about input usage. Even the Japanese commitment to export markets can be viewed as a consequence of the comparative advantages, or disadvantages, of the Japanese steel industry. Lacking a domestic resource base, Japanese steelmakers were forced to export to pay for raw material imports. Needing to be competitive in world markets, they sought the maximum economies of scale—a goal that led to further dependence on exports. The fact that this required some additional expenditures of labor in the packaging, shipping, and overhead categories does not mean that the Japanese steel industry is less efficient, given the markets in which it functions.

It is also true, however, that altered economic choices can have a significant impact on the potential productivity performance of a firm or industry. For that reason, excessive passivity in regard to these types of decisions may represent a lost opportunity. At the very least, the potential benefits of increased scrap consumption should be weighed very seriously even by firms or industries that are accustomed to reliance on hot metal.

Technology

Our discussion of investment presented some aggregate evidence on the link between technology and productivity, using the capital-labor ratio as a proxy for the adoption of new technologies. This discussion can be supplemented by a microeconomic analysis based on the data presented in Chapter 5. We will divide technological differences into two subcategories: major technical changes (i.e., the BOF and continuous casting) and operating techniques, a residual category encompassing myriad incremental improvements in facilities or operating practices. After discussing these two types of technological differences, we will attempt to characterize the overall level of technology in several national steel industries through the proxy of average facility ages.

Major Technical Changes: the Adoption of BOFs and Continuous Casting. The impact of these innovations can be described by focusing on productivity trends within the steelmaking stage, broadly defined to encompass all operations from the steel furnace to the production of the semifinished shapes (slabs, blooms, and billets) which are rolled into finished products. For integrated producers of flat-rolled products, this stage was made up of three discrete steps in the 1950s: the production of steel in open hearths, ingot casting, and slab rolling. By the 1960s, the BOF had replaced the open hearth in this schema, without affecting the other two steps. Continuous slab casting replaces these latter two steps. We can thus define three process routes, denoting different levels of technological progress, within the broadly defined steelmaking stage: open hearth-ingot, BOF-ingot, and BOF-caster.

Figure 6–2 describes productivity trends in each of these process routes, and in the overall steelmaking stage, for both the United States and Japan. As the diagrams illustrate, labor requirements declined over time in all processes due to minor technical improvements and other changes, such as scale increases. Such progress can be viewed as a downward movement along the curves representing each of the process routes. Yet productivity improvements also result from the replacement of one process route by another, that is, shifts between curves from a higher to a lower manhour configuration. Such shifts can be exclusively attributed to technical change.

The effects of such technological discontinuities are clearest for Japan. There the aggregate curve for steelmaking (including slabbing) declines much more rapidly than is the case for any individual set of process steps. Moreover, the gaps separating the processes (i.e., the extent to which a superior process route is in fact superior) are much wid-

Figure 6–2. Technical Changes and Labor Use: Steelmaking.

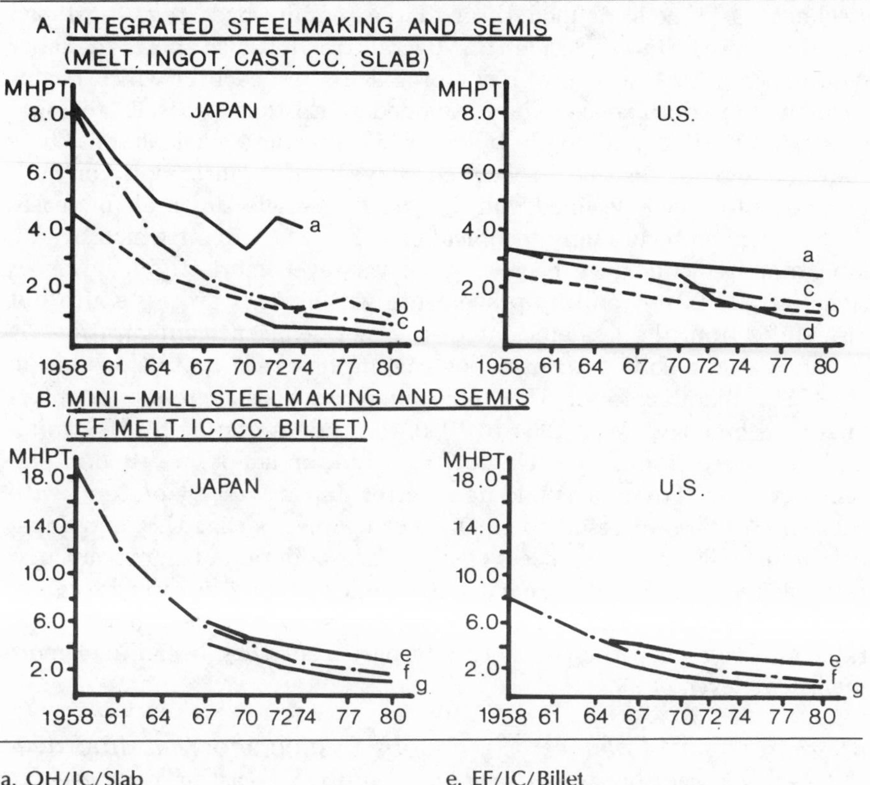

a. OH/IC/Slab
b. BOF/IC/Slab
c. Total (Integrated)
d. BOF/CC (Slab)
e. EF/IC/Billet
f. Total (Mini-Mill)
g. EF/CC (Billet)

Sources: Same as for overall productivity analysis (see Appendix C).

er in Japan than in the United States. This is due to the combination of two factors: the relative efficiency of U.S. open hearths and the relative inefficiency of U.S. continuous casters. Finally, as was discussed in Chapter 3, the U.S. industry has been much slower to shift from one process route to another, especially in regard to continuous casting. More rapid adoption of major new technologies as well as the greater impact of these technologies have combined to provide Japanese integrated producers with a significant productivity advantage.

The reader will probably already suspect that the analysis of major technical changes in the mini-mill sector unearths patterns and results quite different from those found in the integrated sector. As is the case in regard to overall productivity performance, the record of U.S. mini-mill producers in adopting new technologies is quite good, matching if not exceeding the performance of their Japanese counterparts. Figure 6–2 also shows the trend in electric furnace labor requirements for the steelmaking process, defined through the semifinished stage, in Japan and the United States. Besides this overall trend, it also shows the labor requirements for the two different process routes, each of which begins with the electric furnace. The outmoded route then proceeds to ingot casting and primary rolling of billets, while the more efficient route uses continuous billet casting. In each country, labor requirements for both process routes have declined sharply over time—reflecting scale increases that in turn embody many technical changes. The rate at which overall labor requirements have been reduced was even more rapid, however, since it reflects not only improvements within each process route but also shifts from the less efficient to the more efficient configuration.

Whereas the shift to continuous casting has been quite slow within the U.S. integrated sector, U.S. mini-mills have been aggressive adopters of this technology. From 1964 to 1980, continuous casting by U.S. mini-mills increased from 20 percent to 70 percent, an adoption rate of 3 percent per year. The corresponding rate for Japan was 4.4 percent, with the same 70 percent result in 1980. Thus it appears that U.S. and Japanese mini-mills are more or less even in their adoption of significant new technologies, just as they are currently neck and neck in regard to absolute productivity levels. Japanese mini-mills have in fact made more rapid progress than their U.S. counterparts, but they began at a more inefficient level.

Major Technical Changes: Continuous Casting and Yield Improvements. As was shown in Chapter 5, continuous casting has been the principal source of Japanese productivity improvements over the past twenty years, largely because its contributions to improved productivity are multidimensional. Continuous casting boosts labor productivity directly by combining ingot casting and slab rolling. It also raises product quality and therefore reduces the labor time required for upgrading and quality monitoring. Finally, continuous casting has a profound indirect effect on productivity through its potential for increasing yield. Higher yields imply that less raw steel, and thus fewer manhours, are required to produce the finished output.

Because it eliminates the ingot stage, continuous casting can improve yields to the semifinished level by up to 15 percentiles—even more for bar products, since more steps are replaced. Japan's more rapid adoption of continuous casting has provided it with a significant yield advantage and, concomitantly, important benefits in terms of productivity. Japanese integrated producers improved their raw-steel-to-slab yield, reducing raw steel requirements, from 1.16 in 1958 to 1.06 in 1980—largely because of continuous casting. Had U.S. integrated producers achieved this yield in 1980, this would have reduced labor requirements for cold-rolled sheet by 5 percent, all else being equal. U.S. mini-mills, on the other hand, have matched their Japanese counterparts in terms of yield, although the rate of progress has again been greater in Japan. Through its effects on yield alone, continuous casting contributed 7.5 percent of the total productivity improvement in Japanese mini-mills from 1958 to 1980; the equivalent figure for the United States was 3 percent.

For integrated producers, the Japanese benefit not only from more extensive use of continuous casting but from greater efficiency in using it. Continuous casting is a finicky process. It requires extremely tight controls to guarantee sequential pouring and continuous operation. Sporadic operation is the bane of efficient continuous casting: during downtimes the caster can change in size as a result of sharp changes in heat, steel debris can accumulate, and so on. Continuous monitoring, anticipation of developing problems, and a regular supply of molten steel are essential if a caster is to operate smoothly. In this regard, it is highly significant that in 1980, indirect labor requirements in continuous casting operations were 0.36 hours in the United States and 0.12 hours in Japan. Indirect labor requirements basically refer to maintenance operations, so that this discrepancy indicates the extent to which Japanese casting practice is superior to that prevalent in the United States. Better casting practice could be treated as a difference in operating techniques, but we would contend that it should be ascribed to Japan's more thorough adoption of this major technological innovation.

Operating Techniques. Besides the major technical advances that have been described above, a wealth of incremental improvements have contributed to productivity growth in both the American and Japanese steel industries. We refer to such improvements as operating techniques. Unfortunately, it is impossible to directly quantify or portray the impact of such changes on performance. Changes in operating techniques are simply too numerous, and their effects too complex, to permit this.

Three facets of improved operating techniques are worth discussing briefly: changes in equipment, changes in operating practices, and the application of computer controls. These categories clearly overlap; changes in equipment, for instance, inevitably entail changes in operating practices. Newer facilities generally make it possible to attain productivity improvements even if the function involved has not experienced a major technical breakthrough. This type of technical change has been particularly characteristic of rolling mills and finishing operations. To some extent, the introduction of fully continuous hot strip mills in the late 1950s represented a major technological improvement, even though its effects on productivity cannot be quantified. Continuous annealing may also prove to be a breakthrough technology. But such innovations have been supplemented by a steady stream of refinements—often associated with increasing scale—in rolling mill technology: better gauge controls, faster rolling speeds, more continuous processes, and so on. Together, such marginal improvements in equipment provide extremely significant benefits.

The second category of operating techniques involves the operating practices according to which the work process is organized; improved operating practices generally do not entail major capital expenditures. Instead, better operating practices are usually the product of improved industrial relations, better training, and the consistent, purposeful refinement of procedures. They include such innovations as ladle metallurgy, bottom pouring of ingots, ingot tailoring, more efficient use of inputs, and so on. Coke rates—i.e., the ratio of coke required to produce a quantity of iron—provide one example of how operating practices, for example, adjustment of the burden mix or higher pressures in the blast furnace, can affect productivity—although improved equipment and greater scale also lower coke rates. From 1958 to 1980, the Japanese coke rate fell from 0.67 to 0.45—an improvement of 33 percent. In the U.S., the coke rate fell from 0.8 to 0.57—an improvement of 29 percent.[6] Lower coke rates reduce input requirements and thus indirectly reduce labor requirements as well. Minor refinements that affect coke rates therefore offer the potential of improving productivity as well as lowering raw material and capital costs.

A third major source of improved operating techniques involves the general application of data processing and process controls. Computer controls make it possible to monitor performance precisely—a potential that directly boosts both productivity and quality. Insofar as increased quality reduces reject rates, this also provides indirect productivity benefits akin to those generated by yield improvements.

Data processing technology is used more extensively by the Japanese steel industry than by its American counterpart.[7] In the United States, computer controls are generally applied in specific processes (e.g., new blast furnaces) or for specific functions (e.g., keeping track of customers' orders); data processing is added on to existing operations. In Japan, computerization tends to be built into the entire flow of production, providing significant synergistic benefits—e.g., in facilitating a more continuous production process. This manifests itself in a higher quality product. Comprehensive computerization also makes it possible to constantly monitor the overall production process—a capability that is a prerequisite for maintaining a rapid tempo of improvement.

Improved yields provide some quantitative indication of the cumulative importance of improved operating techniques. While yield improvements from raw steel to semifinished shapes can be ascribed to continuous casting, yield improvements from the semifinished to the finished product are due to the cumulative effects of numerous refinements in operating techniques. For Japanese integrated firms, yield improvements at this stage have actually exceeded those achieved from continuous casting; from 1958 to 1980, Japan's yield from slabs to cold-rolled sheet increased from 0.79 to 0.91—a reduction in yield losses of 57 percent. Over the same period, the corresponding improvement in yield losses in the U.S. industry was a miniscule 6 percent (slab-to-cold-rolled-sheet yield changed from 0.82 to 0.83). In mini-mills, the Japanese yield from billets to rods increased from 0.88 to 0.95—a reduction in yield losses of 58 percent. U.S. mini-mills increased billet-to-rod yields from 0.91 to 0.94—a 33 percent reduction in yield losses. These results suggest that U.S. integrated firms are not attuned to seeking continuous marginal improvements in operating techniques, even though the effects of such a commitment can be highly significant. The Japanese are continually redefining the standard of acceptable yield results even in processes that have not undergone major technical changes, and the same can be said of U.S. mini-mills.

The Age of Facilities. Technical changes are frequently embodied in new facilities. Generally, the newer a facility, the more advanced its technology, so that facility age is a suitable proxy for the level of technology. In a previous section of this chapter, the value of the capital stock was related to capacity to provide a rough indication of the relative age of U.S. and Japanese steel facilities. Those relationships suggested that U.S. facilities are now slightly less than twice as old as those in Japan. This is confirmed by the microeconomic data presented in

Table 6–2. Estimated Average Ages of Steel Facilities: U.S., Japan, and Canada[a]

Facility	U.S. Mean	U.S. Median	Canada Mean	Canada Median	Japan[b] Mean	Japan[b] Median
Coke Ovens	17.0	13.0	14.4	13.0	11.4	10.7
All steelmaking[c]	14.0	13.1	13.1	13.3	12.1	10.4
Open hearth	30.0	28.0	28.6	29.0	NA	NA
BOF	13.3	13.8	11.2	14.0	11.3	10.4
Electric furnace	10.7	9.8	9.3	6.4	12.5	10.3
Primary rolling & continuous casting[c]	23.0	22.5	11.3	8.8	10.1	8.0
Primary rolling	27.5	27.0	14.1	15.0	13.7	11.4
Continuous casting	7.2	5.9	6.5	6.0	6.8	6.1
Rolling mills[c]						
Plate mills	23.0	17.0	16.1	16.1	13.1	11.5
Hot-Strip mills	19.3	16.3	18.0	19.0	15.5	13.8
Cold-Strip mills	22.0	21.5	17.5	17.2	13.6	12.8
Wire-Rod mills	14.6	13.0	12.9	17.0	12.7	11.1
Bar mills	20.0	11.0	11.8	7.2	14.7	13.1
Galvanizing lines	19.5	18.5	15.8	21.0	12.6	12.0
Aggregate[c]	18.8	17.0	13.4	14.0	11.7	10.0

a. As of January 1, 1982, counting 1981 as year zero.

b. 1979 data extrapolated to 1982 based on data supplied by one (major) firm.

c. All aggregates are weighted by contribution to total capacity of facilities concerned. For the calculation of the overall aggregate, steelmaking, rolling of semi-finished shapes, and rolling of finished products are weighted equally; coke ovens are weighted according to input/output relationships with steelmaking.

Sources: 33 Metal Producing Magazine, *World Steel Industry Data Handbook, Volumes 2, 4,* and *5* (New York: McGraw-Hill, 1978, 1981, and 1982, respectively); Institute for Iron and Steel Studies (IISS), *The Steel Industry in Brief: Databook, U.S.A.; 1979-1980* (Green Brook, New Jersey: IISS, 1979); American Iron and Steel Institute (AISI), *Iron and Steel Works Directory of the United States and Canada* (Washington and New York: AISI, various years). Data were also provided by Canadian and Japanese steel firms.

Table 6–2, which shows the authors' estimates of the weighted capacity age of various facilities in the United States, Japan, and Canada as of January 1982.

U.S. facilities are on average much older than Japanese and thus embody older technology.8 The mean age of all Canadian facilities is closer to the Japanese level than to that of the United States. In general, the age of Japanese equipment is much more uniform than is the case for both the United States and Canada—a fact that reflects the thoroughness of Japanese expansion and modernization versus the uneven pattern of U.S. investment. Insofar as relatively modern facilities are linked with older ones, the net benefit for product productivity is slight. The Japanese industry has been able to undertake a broad program of modernization, so that there are few technological bottlenecks. The U.S. industry, on the other hand, has been able to undertake relatively few major projects, so that the unbalanced character of U.S. modernization has significantly reduced the efficiency of investment and the extent to which up-to-date technologies are used. Older facilities imply older technology and this in turn implies lower productivity; the link among these

Table 6–3. Ranking of Productivity vs. Mean Age: U.S., Japan, and Canada.[a]

	U.S.		*Canada*		*Japan*	
Facility	*Age*	*MHPT*	*Age*	*MHPT*	*Age*	*MHPT*
Coke oven	3	3	2	2	1	1
Steelmaking						
Open hearth	2	2	1	1	NA	
BOF	3	3	1	3	1	1
Electric furnace	1	1	1	1	3	3
Rolling mills						
Plate	3	3	2	2	1	1
Hot strip	3	3	2	2	1	1
Cold strip	3	2	2	3	1	1
Wire rod	3	3	1	1	2	2
Bar	3	3	2	2	1	1
Galvanizing	3	3	2	2	1	1
Average rank	2.7	2.5	1.5	1.8	1.3	1.3

a. 1980 process productivity; 1982 facility ages.
Sources: Tables 6–2, Table 5–6, and data provided by Canadian steel firms.

variables is suggested by Table 6–3. Ranking U.S., Japanese, and Canadian facilities by age gives almost the same result as a ranking by productivity.

Table 6–2 can also be used to describe investment patterns. Low mean ages are associated with some of the facilities that have been the targets of U.S. modernization efforts in recent years: BOFs, electric furnaces, continuous casters, and wire rod mills. A substantial difference between mean and median age is another indicator that a facility has been a recent investment target. This differential is greatest for bar mills, where it reflects the expansion of the mini-mill sector, and for coke ovens, where it reflects the consequences of new construction and retirement. Environmental regulations have encouraged the industry to devote substantial sums to modernizing coke batteries, although the productivity potential of coke oven investment is limited. Far greater improvements in efficiency could have been obtained had the funds spent on coke ovens been devoted to investments that embody major technical advances, such as continuous casting. In a sense, coke oven investment was mandated by environmental regulations; yet, substitutes (e.g., purchasing coke or building more electric furnaces to increase scrap use) have not been aggressively sought. Instead, steel firms have often adhered to the goal of self-sufficiency in processes like coke production in order to maintain their vertically integrated structure.

The relatively advanced age of U.S. steel facilities and the uneven pattern of modernization (both shown in Table 6–2) reflect the technological basis for the poor productivity performance of the U.S. industry relative to its Japanese competition. It is difficult to quantify the effects of technology, yet the various themes that have been discussed in this section support the view that technological differences have been significant determinants of the comparative productivity performance of the U.S. and Japanese steel industries. Just how significant will be described in the next chapter.

Scale

In relating the observed record of productivity growth to the scale of various facilities in the U.S. and Japanese steel industries, we will define a facility's scale as its capacity, or output, per unit of time. Defined in this way, scale is a complex rather than self-evident concept and acts as a residual category in our analysis. Any increase in efficiency that boosts the throughput capacity of a facility can manifest itself as an increase in scale–regardless of the reason for the increase.

Figure 6–3. Scale and Productivity: Selected Integrated Facilities: 1958–1980.[a]

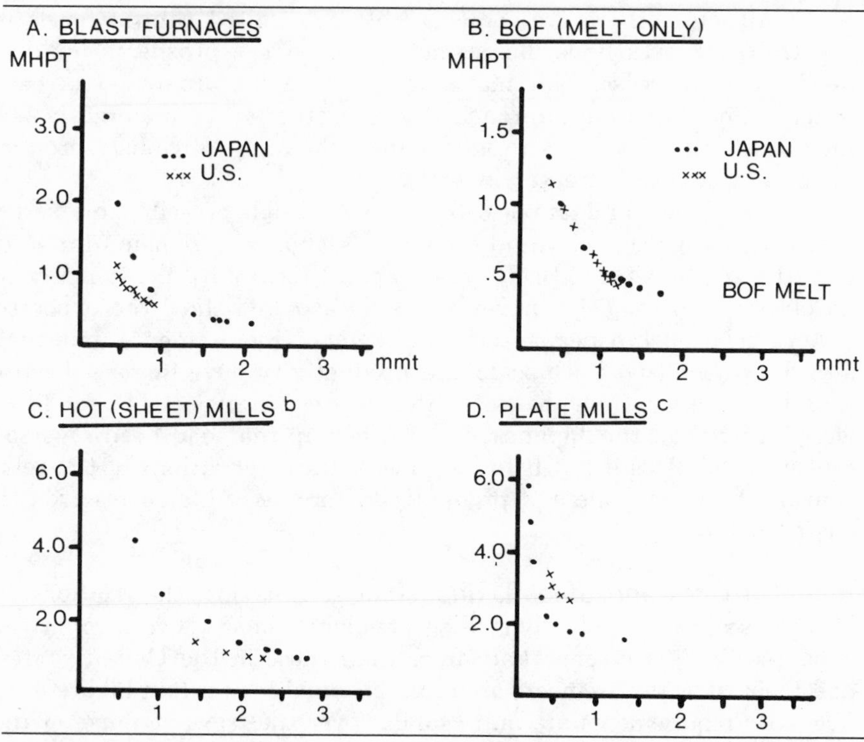

a. 1958, 61, 64, 67, 70, 72, 74, 77, 80 except for US hot (sheet) mills and plate mills (1958, 64, 72 and 80).

b. Includes condition.

c. Includes finishing.

Sources: Productivity as in Chapter 5 (see Appendix C); capacity data from American Iron and Steel Institute (AISI), *Directory of Iron and Steel Works in the United States and Canada* (New York and Washington, D.C.: AISI various years), U.S. Bureau of the Census, *Census of Manufactures* (Washington, D.C.: U.S. Government Printing Office, various years); Japan Iron and Steel Federation (JISF), *Statistics of the Iron and Steel Industry* (Tokyo: JISF various years) and *Statistical Yearbook* (Tokyo: JISF, various years).

Scale increases in the Japanese steel industry have represented progress on all fronts: greater size, the adoption of major innovations, and improved operating techniques. Scale-related productivity improvements in the United States, however, have more often been based on technological changes or altered operating practices than on increases in the actual size of facilities. Faced with capital constraints and a slow-growth market, the U.S. industry has been less able to make great strides in regard to the size of its facilities. Marginal changes (e.g., with blast furnaces: higher pressures, oil injection, and better tops) have been the most common route to increased scale. Although there are some exceptions, limited capital funds in the United States have increasingly reduced the prospects for dramatic scale increases, so that more incremental progress has been sought. The Japanese approach—the construction of mammoth new facilities built ahead of a rapidly growing market—obviously produces more dramatic increases in scale.

This is shown in Figures 6–3 and 6–4, which present time series data on average capacity and on average labor requirements for steel facilities in the United States and Japan.[9] Each data point refers to an observation for a given year. The data points show the expected inverse relationship between scale and labor requirements, but they also document that both scale and productivity have increased more rapidly in Japan than in the United States. In 1958, U.S. facilities were larger than the Japanese—a relationship that had been reversed by the early 1970s if not before. Some of the implications of the relationship between scale and productivity for specific facilities are discussed below.

Blast Furnaces and Steelmaking. For steelmaking, the relationship between scale and productivity is surprisingly similar between countries. Japanese BOFs are somewhat larger than those in the United States, and their labor requirements are correspondingly lower (see Figure 6–3). Yet labor requirements are quite similar over the relevant range of the curves; for observations where the size of facilities are more or less the same, labor requirements are also approximately identical. This is even more apparent in regard to electric furnaces (Figure 6–4), where both scale and productivity are presently very close. This similarity in performance reinforces the view that productivity is principally determined by technical factors. While the above observations are based on serial data, Japanese cross-sectional data, describing 1978–80 labor requirements for different scales, show the same relationship between scale and labor requirements in the BOF. This suggests that major technical changes have had little impact in raising BOF scale. For electric furnaces, the cross-sectional relationship is somewhat flatter than the one described by serial data (Figure 6–4), suggesting a greater role for technical changes in increasing scale.

These curves indicate that steelmaking labor requirements for U.S. integrated producers would be as low as for their Japanese competitors were the scale of facilities identical. As shown in Figure 6–3, this does not seem to be the case in regard to blast furnaces, where U.S. labor requirements for a given scale seem to be lower than in Japan. (It should be noted that blast furnace data refer to actual output, whereas the scale of other facilities is measured in terms of capacity.) This may be partly due to discontinuities resulting from vast differences in blast furnace scale at a given point in time. The 1980 observation for the United States represents an annual output of 800,000 tons, while Japan achieved this average output level in 1968. The main explanation, however, involves technical change. Using cross-sectional Japanese data (for 1978–80), a flatter relationship between scale and MHPT is evident; it follows the Japanese data points in Figure 6–3 at capacities above one million tons and the U.S. data points for furnaces with lower capacities. Serial and cross-sectional Japanese data thus diverge at annual capacities below one million tons, the average size prior to 1970. This suggests that during the 1960s the Japanese made major strides to catch up to the United States in blast furnace technology, probably through improved input use, and had caught up by the 1970s. The cross-sectional data also suggest that, given the same technology, the relationship between scale and MHPT in blast furnaces is similar in the United States and Japan.

The data presented in these graphs (supplemented by cross-sectional data) can also be used to assess the minimum efficient scale of operations, judged in terms of labor productivity alone. Minimum efficient scale is defined as the point where these curves begin to level off. Referring to the Japanese data, it would seem that the minimum efficient scale of an individual blast furnace is 1.5 million tons of annual output; for a BOF vessel, the minimum efficient scale is 1.5 million tons of annual capacity. This implies that the minimum efficient scale for ironmaking or steelmaking shops is at least twice these levels, since integrated plants generally require two or more blast furnaces or steelmaking furnaces in order to ensure sequential pouring, better caster performance, and the availability of inputs during maintenance downtimes. These estimates of minimum efficient scale, while representative, apply only to labor usage; total costs must be considered before one can define the overall minimum optimal scale (see Chapter 7).

The minimum efficient scale for an electric furnace amounts to less than 200,000 tons per year—only slightly more than 10 percent of the level for a BOF. As is the case with BOFs, this usually implies a new electric furnace shop of at least twice this level, since most plants use at least two furnaces for continuous operation. The relatively limited scale required to attain most of the potential operating efficiencies in electric furnaces has had a profound effect on structural changes in the steel industries of both Japan and the United States. There have traditionally been substantial economies of scale in the primary end of steelmaking, and this has been one of the foundations of the outstanding

Japanese performance from 1955 to 1975. Through most of the 1950s and 1960s, Japanese investment was focused on blast furnaces and steel furnaces, where economies of scale are most significant. Yet the scale advantages of large integrated facilities for the production of raw steel are no longer so definitive. The electric furnace can now hold its own in terms of labor productivity against the integrated process, even at much lower scale. The increasing attractiveness of the electric furnace as a source of carbon steel portends further fundamental realignments in the structure of the steel industry; it has already radically altered the traditional configuration of economies of scale.

Finishing Facilities. Figures 6–3 and 6–4 show that labor productivity increases with scale in finishing facilities as well as at the hot end. Only in the hot strip mill, however, does it appear that high output levels are required to gain most scale benefits at the finishing end. The annual capacity of Japanese hot strip mills averaged almost 3.5 million tons in 1980, yet it is still not clear that the minimum efficient scale has been reached. Somewhat surprisingly, U.S. firms have apparently outperformed their Japanese competitors in hot strip mills of the same size. This may be due to the fact that this is the product on which U.S. firms have traditionally concentrated and for which U.S. producers have traditionally been the world leaders technologically. It may also be due to other factors, such as differences in the gauge and width of the steel rolled and the Japanese commitment to higher quality.

Considering only the Japanese data, which are more extensive, it would appear that minimum efficient scale amounts to about 700,000 tons in plate mills. The discrepancy between the U.S. and Japanese curves in this case is probably due to technological differences, as reflected in the advanced age of U.S. plate mills. In other cases, especially wire rod mills (Figure 6–4), small increases in scale seem to be associated with dramatic productivity improvements. Yet the minimum efficient scale appears to be reached at fairly low levels of output relative to the pattern found in hot strip mills. While the data do not permit the definitive identification of the minimum efficient scale in regard to wire rod mills, a level of about 500,000 tons is plausible. Although not shown in Figure 6–3 or 6–4, the minimum efficient scale is estimated to be roughly 1 million tons for cold-rolling mills, 250,000 tons for seamless tube mills, and 500,000 tons for bar mills. These data explain the attractiveness of barlike products to mini-mills; furthermore, they suggest that efficiencies of scale are not a significant barrier to entry in regard to the production of seamless tubes, plates, and so on.

For finishing facilities, careful consideration of scale-related efficiencies (at least in terms of labor productivity) thus belies the commonly

Figure 6–4. Scale and Productivity: Selected Mini-mill Facilities, 1958–1980

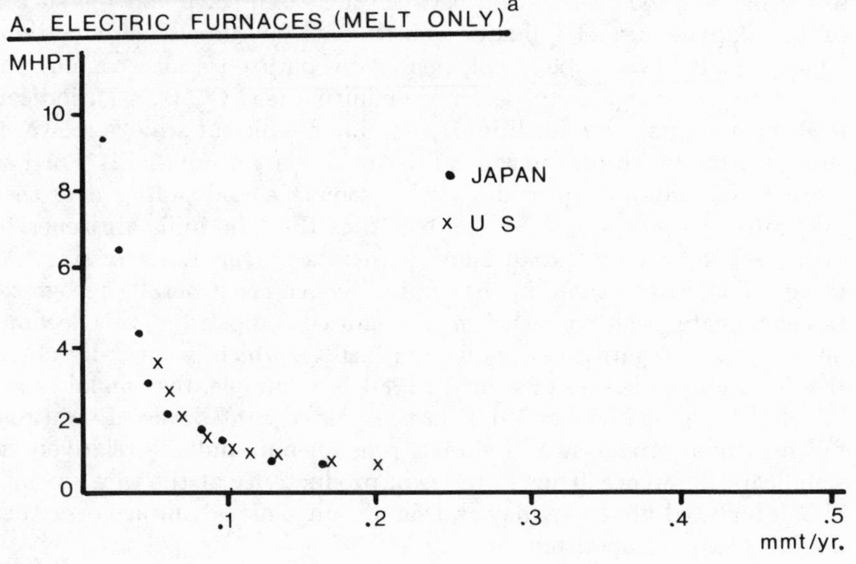

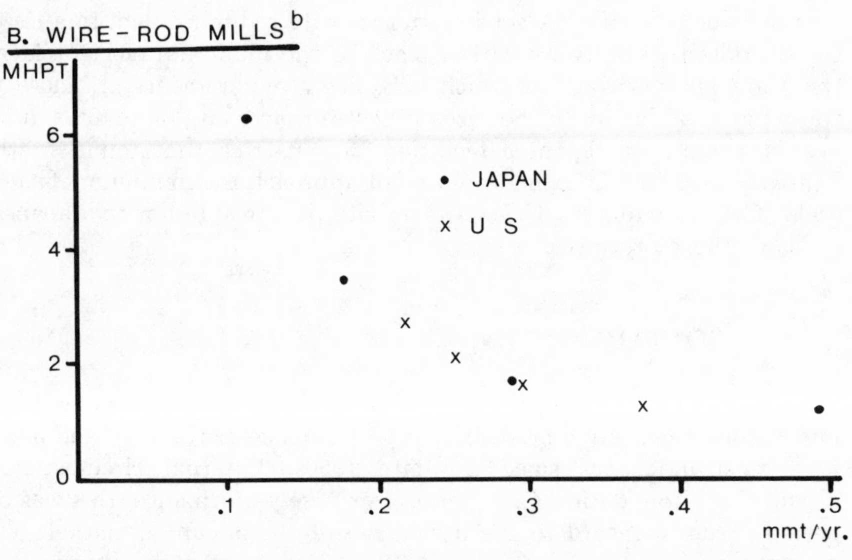

a. 1958, 61, 64, 67, 70, 72, 74, 77, 80
b. 1958, 64, 72, 80. Includes finishing.
Sources: As in Figure 6–3.

held perception that large steel mills enjoy a clear superiority over smaller ones—a perception that has been greatly strengthened by a somewhat spurious correlation between increased scale and the success of the Japanese steel industry. These data on process productivity, which would have to be supplemented by data on scale economies in regard to other inputs to be truly definitive (see Chapter 7), indicate that for most finishing facilities the minimum efficient scale is relatively small—certainly within the capabilities of the larger mini-mills. Furthermore, most mini-mills probably derive some scale advantage over their integrated counterparts from the fact that the mini-mills are generally dedicated to only one product and even to a narrow range of sizes. Although integrated plants in the United States are generally recognized to have greater scale than their nonintegrated competitors, this need not be the case in regard to an individual facility, which is the truly operative level for efficiencies of scale. In 1980, for example, the annual capacity of the average wire rod mill in the United States was 415,000 tons for integrated producers and 371,000 tons for mini-mills—a relatively insignificant difference. Thus in terms of productivity at the wire rod mill, U.S. integrated producers have no significant scale advantage over their nonintegrated competitors.

Overall, these results show that scale is an important determinant of productivity and can explain some of the results presented in Chapter 5. The scale of U.S. BOFs, electric furnaces, wire rod mills, and, to a lesser extent, hot strip mills are within reach of the minimum efficient level; these are all processes for which U.S. labor requirements are close to those found in Japan. In the case of blast furnaces and plate mills, however, the scale of Japanese facilities far outstrips that found in the United States; and U.S. facilities do not approach the minimum efficient scale. Correspondingly, U.S. productivity is far inferior to Japanese levels in those processes.

CONCLUSION

This chapter has described a number of factors that have jointly determined the comparative productivity performance of the U.S. and Japanese steel industries since the late 1950s. External circumstances provided the foundation for superior Japanese performance; this was especially true in regard to the dynamism of the Japanese market. The corporate strategies of Japanese steel firms were well tailored to exploit these favorable circumstances. A spectacular rate of investment was sustained for a long period, and this investment was embodied in ultra-large, modern plants. Between 1960 and 1975, the Japanese revolutionized the world steel industry's perception of economies of scale; and their facilities are now more modern than those of any other major steel industry. As a result, the comparative disadvantage of a poor resource base has been nullified. This convergence of positive factors began to deteriorate for the Japanese steel industry in the 1970s—as evidenced by the slower productivity growth of that decade. Nonetheless, for the period 1958 to 1972, every element of the productivity puzzle seemed to be in place for Japan's steelmakers.

The circumstances faced by the U.S. industry, on the other hand, have been more adverse. Judged in terms of comparative statics, U.S. producers in the 1950s enjoyed a commanding advantage over their Japanese competitors, yet they have been unable to implement strategies which could generate dynamic cost reductions within a slow-growth environment. Weak market growth placed significant constraints on capital expenditures, effectively barring the U.S. industry from matching Japanese investment. Investments were too widely dispersed, and they often failed to target areas that could provide significant productivity improvements. Low rates of investment have led to lagging implementation of new technologies and only gradual increases in scale. As a result, U.S. facilities are now relatively old and relatively small. This description does not apply to the mini-mills, however; this sector has been the sole bright spot in the generally gloomy picture of steel industry performance.

On every front—labor relations, technological aggressiveness, self-monitoring of performance—integrated firms in the United States are still groping for an effective cost reduction strategy. Japanese producers and U.S. mini-mills are far more proficient at this. How and to what extent domestic integrated firms can make up this gap is the principal theme of the next chapter.

Notes

1. John Stuart Mill, *Principles of Political Economy with Some of Their Applications to Social Philosophy* (Toronto: University of Toronto Press, 1965), p. 101.
2. World Steel Dynamics, "Steel Strategist #6" (New York: Paine Webber Mitchell Hutchins, 1982), Table 5.
3. Calculated from American Iron and Steel Institute (AISI), *Annual Statistical Report, 1981* (Washington, D.C.: AISI, 1982), p. 10.
4. Given the assumptions used for these calculations (twenty-year, straight line depreciation), annual investment must amount to 10 percent of the value of the capital stock in order to maintain that

value. In 1980, for example, this would have necessitated capital expenditures of roughly $4 billion.

5. AISI, *Annual Statistical Report, 1981*, p. 13a.
6. Japan Iron and Steel Federation (JISF), *Statistical Yearbook for 1960* (Tokyo: JISF, 1961), pp. 1 and 18; S. Kawata, ed., *Japan's Iron and Steel Industry, 1981* (Tokyo: Kawata Publicity, Inc., 1981), p. 52; AISI, *Annual Statistical Report, 1958* (New York: AISI, 1959), p. 52, and *Annual Statistical Report, 1980* (Washington, D.C.: AISI, 1981), p. 64.
7. See, for example, Bela Gold, "Factors Stimulating Technological Progress in Japanese Industries: The Case of Computerization in Steel," *Quarterly Review of Economics and Business* 18, no. 4 (Winter, 1978): 7–21.
8. These data may overstate the age difference, since our calculations consider the effects of major modernization projects but not the impact of incremental improvements, which are more characteristic of the U.S. industry.
9. More extensive historical data on facility capacity are presented in Table 3–4, while comparable data on productivity by facility are presented in Table 5–6.

7 IMPROVING PERFORMANCE: WAYS AND MEANS

It is a joke in Britain to say that the War Office is always preparing for the last war . . .[1]

—*Winston Churchill*

The superior productivity performance of the Japanese steel industry is an accomplished fact, one that cannot be ignored by U.S. firms, however uncomfortable it may be. The Japanese performance has been based on a very solid foundation; rapid market growth has fostered robust investment, embodied in massive state-of-the-art plants. The competitiveness of U.S. integrated producers, by contrast, has been deteriorating; and most of the evidence points to continued and even accelerated deterioration in the status of the U.S. industry. Major investments seem to be required to raise its performance to the Japanese level, yet the industry's desperate need for more modern facilities is thwarted by the difficulties it has raising the funds its modernization plans require.

The link between performance and investment confronts the U.S. steel industry with a seemingly intractable dilemma. Lagging performance reduces the industry's competitiveness and profitability. Physical efficiency could be improved by investment in larger, more up-to-date facilities, but experience indicates that such investments may not improve profitability. As a result, the industry has remained an unattractive outlet for investment; meaningful performance improvements therefore seem perpetually out of reach.

This dilemma defines the industry's principal strategic problem, and it is also the chief theme of this chapter. Our analysis will begin by evaluating the investment strategy that has prevailed within the industry; we will call this the "massive-modernization" strategy, a legacy of the era in which domestic integrated producers dominated the U.S. steel market. Productivity data will be combined with estimates of other costs, especially unit capital costs, to assess the economic viability of this strategic approach. After identifying some of the problems inherent in the massive-modernization strategy, we will reexamine the data presented in Chapters 5 and 6 to determine which of the immediate determinants of performance has the greatest impact on productivity and on overall cost performance. An assessment of the relative importance of technology and scale is especially revealing, and this issue constitutes the middle section of this chapter. Finally, we will use this analysis to suggest the outlines of a more viable alternative to massive modernization.

THE MASSIVE-MODERNIZATION STRATEGY

From at least the time of Judge Gary and Andrew Carnegie, there has been one predominant view of the optimal way to boost productivity performance: construct large, complex, greenfield plants, using proven new technology, producing a wide range of products, and built to last. The postwar experience of the Japanese steel industry has tended to confirm this view, even though lagging markets, inadequate funds, and other constraints have forestalled this option for the U.S. steel industry, at least since the mid-1960s.

Because of capital constraints and market limitations, there is now little if any talk of greenfield (new site) investment in integrated steel production in the United States. Nevertheless, the greenfield ideal continues to define the massive programs that integrated producers view as the key to maintaining or regaining competitiveness. When this kind of thinking is applied to the revitalization of existing plants, it is referred to as "brownfield" modernization. Its goal is the gradual transformation of an existing site into a sort of poor man's greenfield plant, replacing older facilities with larger and more up-to-date units. In the United States, this greenfield strategy now appears only in its brownfield variant. Its underlying logic, however, is the same: improved performance is dependent on massive expenditures for facilities that are large, capital intensive, modern, and durable.

Clearly, even the brownfield variant of the massive-modernization strategy is impossible to implement unless large sums are available for investment. The industry's inability to raise funds on such a scale thus generally prohibits it from implementing this strategy for improving performance, although there are several examples of enormously expensive partial implementation, such as the large blast furnaces built in the late 1970s at Inland's Indiana Harbor Works and Bethlehem's Sparrows Point plant. Rather than forsake its preferred modernization strategy, the industry has expended enormous efforts to boost the sums available to it, particularly by seeking tax relief and reductions in mandated expenditures. In January 1980, the American Iron and Steel Institute released a program for revitalizing the industry, predicated on estimated annual investment requirements of over $8 billion in 1982 dollars. Since that time, steel industry investment has not reached even half of this level—and the steel depression of 1982 does not bode well for future investment.[2] Retrenchment and rationalization, rather than expansion, are now being cited as necessary components of the industry's strategy for improving performance. Yet even the underutilization of the industry's most modern facilities has not altered the view that some variant of the massive-modernization strategy is the key to future success.

The Potential: Productivity Benefits

A greenfield plant holds the promise of dramatic technical improvements in the efficiency with which all inputs are used, and this is the foundation of the massive-modernization strategy's appeal. Although greenfield conditions are presently unattainable for integrated producers in the United States, much of their potential could be realized in an existing plant via the brownfield route of approximating greenfield operations. Just how great a difference this would make can be estimated by using new-plant productivity estimates provided by steel companies and plant builders. These data, which will be discussed in more detail below, refer to both integrated and mini-mill operations and identify the minimum optimal scale for each type of plant. The increased labor productivity described by these estimates is merely one example, although arguably the most important, of the efficiency advantages attainable in a new plant.

In order to produce cold-rolled sheet in an integrated plant, a new mill should have an annual capacity of at least four million tons. Such a plant would minimally comprise two blast furnaces, two BOFs, two slab casters (with no ingot casting), one hot strip mill, and various finishing mills (a combination pickling-tandem line, continuous annealing, etc.). This plant's potential productivity performance is described in Table 7–1. Compared with the average present performance described in Table 5–1, this state-of-the-art plant could lower labor requirements by 40 percent in the production of cold-rolled sheet. Similar labor savings would be realized in other products. For example, the labor required for the production of a ton of wire rods would fall 45 percent—from 6.45 manhours per ton (MHPT) to 3.55 MHPT. Although not shown, state-of-the-art operations would also boost the efficiency with which other inputs are used; energy costs in the production of cold-rolled sheet, for instance, would be reduced by over 20 percent in a new integrated mill.

Table 7–1. Manhours per Net Ton, by Process, in New U.S. Plants (90% operating rate).

	Process				Input/Output	Cumulative	
	Direct		Total			Direct	Total
A. *Integrated: CRS*							
Ore handling				0.04			
Sintering		0.02		0.15			
Coal handling				0.04			
Coke ovens		0.25		0.48		0.25	0.54
Ore/HM					1.45		
Coal/HM					0.65		
Sinter/HM					0.15		
Coke/HM					0.45		
Blast furnaces		0.15		0.36		0.26	0.68
HM/CS					.80		
Scrap/CS					.28		
BOF:							
Melt	0.13		0.40				
Continuous casting	0.10	0.23	0.35	0.75		0.44	1.30
Slab/HR					1.05		
Hot mill	0.18		0.42				
Conditional, Trim	0.14	0.31	0.31	0.74		0.78	2.10
HR/CR					1.06		
Picke	0.13		0.16				
Cold reduction	0.15		0.18				
Anneal	0.15	0.92	0.19	1.19		1.75	3.42
Temper	0.17		0.23				
Finish	0.32		0.43				
Package & shipping			0.60				4.02
Administration & overhead			0.28				4.30
Total							4.30
B. *Mini-mills: Wire rod*							
Electric furnace							
Melt	0.30		0.45				
Continuous casting	0.20	0.50	0.28	0.73		0.50	0.73
Billet/Rod					1.06		
Rod mill		0.48		0.78		1.01	1.55
Package & shipping				0.30			1.85
Administration & overhead				0.15			2.00
Total							2.00

Sources: Authors' estimates based on Japan Ministry of Labor (JML), *Labor Productivity Statistics Survey* (Tokyo: JML, various years) and data provided by steel firms. See note 3 of this chapter and Appendix C.

Although it would resemble present plants, a state-of-the-art integrated mill would have an improved flow of materials and larger, more efficient facilities. Table 7–2 identifies the sources of the advantages enjoyed by state-of-the-art plants producing cold-rolled sheet relative to current practice. Greater use of continuous casting, with improved casting practice, and the resultant higher yields would contribute 38 percent of the total productivity improvement. Better primary facilities would contribute 17 percent, and superior cold

Table 7–2. Sources of New-Plant Productivity Advantages over Current U.S. Plants.

	MHPT	*Share (%)*
A. *Integrated CRS*		
Primary process		
Sinter & material handling	0.06	
Coke ovens	0.16	17
Blast furnaces	0.28	
Input use	0.10	3
Steelmaking (Non-CC)		
Better performance	0.08	
More BOFs	0.12	7
Continuous casting (CC)		
Better performance	0.27	
More CC	0.39	38
Yields	0.43	
Hot rolling	0.20	7
Cold finishing	0.38	13
Package & overhead	0.43	15
Total	2.90	100
B. *Mini-mills: Wire rod*		
Steelmaking	0.27	18
Continuous casting		
Better performance	0.16	
More CC	0.29	33
Yields	0.05	
Rod mill	0.47	31
Package & overhead	0.27	18
Total	1.51	100

Source: Derived from Table 7–1.

finishing and administrative practices would contribute 13 percent and 15 percent respectively. Smaller productivity gains would be realized in steelmaking and hot rolling, where present practice is relatively sound.

State-of-the-art facilities provide significant efficiency advantages in mini-mills as well, and these are also described in Table 7–1. The mini-mill configuration of processes is characterized by a much lower minimum efficient scale; a mini-mill producing wire rods needs to have a capacity of 750,000 tons or less rather than the approximately four million tons required for the integrated production of cold-rolled sheet. At this scale, a state-of-the-art mini-mill would comprise two or three electric furnaces, a like number of continuous billet casters, and one wire rod mill. In such a plant, wire rod production would require only two MHPT versus the present average of 3.51—a productivity improvement of about 45 percent. The sources of improved productivity in state-of-the-art mini-mills are also described in Table 7–2. Greater continuous casting and improved yields are the principal contributors to better performance (33% of the total gain), while better electric furnace melting, via faster heat times and so on, provides 18 percent of the total productivity improvement. Improved rolling practices (31% of total) and better control of overhead expenses (18% of total) also contribute to state-of-the-art advantages.

In mini-mill plants, the transition to state of the art entails roughly comparable progress in each component step. In the integrated case, however, progress is much more uneven: continuous casting provides almost three times the benefits of superior finishing operations. This discrepancy reflects the consistency of mini-mill advances in recent years compared to the more fitful progress achieved by integrated producers. On the whole, existing mini-mills are much closer to the configuration described as state of the art than are their integrated competitors. Indeed, technical refinements could reduce labor requirements to two MHPT in some existing mini-mills, and state-of-the-art performance is being redefined at a rapid pace. No integrated U.S. plant is within reach of state-of-the-art performance, but several mini-mills will probably surpass it in the not too distant future.

Were it possible to wish away the existing steel industry of the United States and replace it with fully modern, optimally sized plants, labor productivity would be boosted by 40 to 45 percent. In integrated products, this would eliminate the Japanese advantage, so that U.S. labor productivity would again set or match the international standard. The data presented in Table 7–1 explain why the greenfield plant is the promised land of productivity performance for integrated producers.

The Pitfalls: Capital Costs

The productivity benefits described in Table 7–1 would stem primarily from the substitution of capital for labor—capital-deepening investment, reflecting technological progress. Yet improved productivity has a price; capital costs rise as capital is substituted for labor. Improving labor productivity, while crucial, is only a means to an end for a company. That end is improved competitiveness overall, measured by the market standard of costs (and prices). If the substitution of capital for labor improves overall performance and competitiveness, a gain is made, but if better performance is offset by increased capital costs, the net result is negative. As Chapter 3 has shown, the American steel industry has a significant cost disadvantage to erase if it wishes to be truly competitive with the Japanese. It is therefore necessary to assess the capital cost effects of massive modernization in order to evaluate the overall viabili-

Table 7–3. Capital Costs per Net Ton by Process in New Plants: Integrated, Cold-rolled Sheet[a] (1981 $, 90% operating rate).

		Process	*Input/Output*	*Cumulative*
Sintering		89		
Coke ovens		305		
Sinter/HM			0.15	
Coke/HM			0.45	
Blast furnaces		200		351
HM/CS			0.80	
BOF:				
Melt	90			
Slab cast	102	192		473
Slab/HM			1.05	
Hot mill		218		714
HR/CR			1.06	
Pickle	122			
Cold reduction	102			
Anneal	89	448		1205
Temper	67			
Finish	28			
Total				1205

a. Overhead (e.g., office buildings) and infrastructure (e.g., plumbing, electrical, etc.) allocated to all facilities. Ore and coal handling allocated to using facilities (sinter, coke ovens, and blast furnaces). Site preparation and financing costs during construction not included.
Source: See Appendix C and note 3 in this chapter.

Table 7–4. Capital Costs per Net Ton by Process in New Plants: Wire Rods[a] (1981 $, 90% operating rate).

		Process	Input/Output	Cumulative
A. *Integrated Plants*				
Sintering		89		
Coke ovens		305		
Sinter/HM			0.15	
Coke/HM			0.45	
Blast furnaces		200		351
HM/CS			0.80	
BOF:				
Melt	90			
Billet cast	48	138		419
Billet/Rod			1.065	
Rod mill		179		625
Total				625
B. *Mini-mills*				
Electric furnace				
Melt	98			
Billet cast	38	136		136
Billet/Rod			1.05	
Rod mill		157		300
Total				300

a. Overhead (e.g., office buildings) and infrastructure (e.g., plumbing, electrical, etc.) allocated to all facilities. Ore and coal handling allocated to using facilities (sinter, coke ovens, and blast furnaces). Site preparation and financing costs during construction not included. Both plants have 750,000 tons of wire rod capacity. Mini-mill plant capacity 750,000 tons, integrated plant capacity 4 million tons.
Source: See Appendix C and note 3 in this chapter.

ty of this strategy—even though conceptual difficulties and the paucity of reliable data prohibit us from developing the sort of international, intertemporal comparisons that we have constructed for labor productivity. Engineering estimates, provided by steel firms and suppliers, have been combined with input-output relationships—similar to those used to measure labor productivity by product in Chapter 5—to develop the capital cost estimates presented in Tables 7–3 and 7–4.[3] The first table refers to integrated production of cold-rolled sheet;[4] the second concerns integrated and mini-mill production of wire rods.

These tables provide several insights into the nature of the capital constraints faced by steel firms. For integrated processes, the high cost of coke ovens and, to a lesser extent, blast furnaces (including sintering) stand out. This highlights the necessity of improving coke rates, raising yields, and lowering hot metal ratios in order to reduce the impact of these capital costs on the total. Second, the data show that cold-finishing equipment is also very expensive. This partially explains why such facilities are so old in the United States. Altering economic choices or operating techniques could reduce the impact of cold finishing on total capital costs. For instance, cold-finishing scale economies could be increased, thus reducing unit capital costs, by simplifying product mix or by combining various process steps into a more continuous flow. Farming out these operations would also be a potential source of capital cost savings.

The mini-mill data presented in Table 7–4 refer to a 750,000-ton plant and are based upon more extensive experience in recent construction. Table 7–4 also estimates the unit capital cost of incorporating 750,000 tons of wire rod capacity into a new integrated plant. In this case, costs were estimated using the assumptions most favorable to integrated techniques.[5]

Even then, however, mini-mills still enjoy far lower capital costs. To a great extent, this is due to the simpler configuration of mini-mills, which are not handicapped by the high cost of primary processes. (Coke ovens, blast furnaces, etc. add about $300 per ton to finished product capital costs.) This advantage is exacerbated, however, by differences in how mini-mills and integrated plants are built and operated—particularly the "build tight, build quick, and build cheap" philosophy described in Chapter 4. The most successful mini-mills are ruthless cost cutters, paring construction costs to the absolute minimum. Furthermore, mini-mills are careful to ensure that the product lines in which they choose to compete will provide revenues that exceed the cost of production; they will not strive for a full range of products and sizes if seeking the last 5 percent of the market adds 25 percent to capital costs. When combined with the inherent simplicity of mini-mill techniques, this philosophy generates significant cost savings for mini-mills. This is reflected in the fact that capital cost estimates are lower for a given facility (e.g., a billet caster or a rod mill) if it is placed in a mini-mill than if it is placed in an integrated plant. This philosophical difference, however, has other effects which are not reflected in Table 7–4, even though they tend to increase the actual mini-mill advantage in capital costs. Profitable mini-mills are characteristically designed to produce a narrow range of products and product sizes, with an overwhelming dedication to meeting or beating competitors' costs. Table 7–4 estimates capital costs for integrated wire rod production as if a similar commitment to minimizing costs were applied. This understates the costs associated with the traditional integrated approach to planning, building, and operating new plants—that is, massive modernization.

The Consequences

The capital cost estimates presented above can be combined with the productivity results and other data to construct a series of cost esti-

Table 7–5. Estimated Costs of New Versus Current (1981) Plants: Integrated and Mini-mills (U.S. $/net ton, 90% operating rate.)

	Current Plants			*New Plants*		
	Integrated		*Mini-mill*	*Integrated*		*Mini-mill*
	CRS	*Wire Rod*	*Wire Rod*	*CRS*[a]	*Wire Rod*[b]	*Wire Rod*[c]
Labor	139	127	59	86	73	35
Iron ore	62	61	—	62	60	—
Scrap	15	14	93	15	14	92
Coal or coke	53	51	—	42	39	—
Other energy	53	45	45	41	35	40
Other costs	82	65	65	70	55	58
Operating costs	404	363	262	316	276	255
Depreciation[d]	16	10	10	92	48	21
Interest[e]	6	4	7	72	38	18
Taxes	6	4	3	12	8	4
Total costs[f]	432	381	282	492	370	268
List prices	486	419	419	486	419	419

a. 4-million-ton plant.
b. 4-million-ton plant, producing bar, rod and light structural 0.75 million tons wire rod.
c. 0.75 million ton plant, all wire rod.
d. New plant capital costs (Table 7–3 and Table 7–4) amortized over fifteen years. Depreciation also includes amortized construction interest—$12 a ton CRS, $6 a ton WR integrated, $1 a ton wire rod mini-mill. This reflects 5-year construction period integrated, 2-year construction period mini-mill.
e. 50 percent debt at 12 percent interest assumed, 50 percent equity.
f. Excluding any return on equity. For the same return on equity, new plants must earn more absolutely than current plants, and new integrated plants must earn more than new mini-mills.
Sources: Authors' estimates, based upon data from: World Steel Dynamics, *Core Reports J & Q* (New York: Paine Webber, 1979 and 1982); *Tex Reports* (Tokyo: various years); *TPM Cost Manual* (Washington, D.C.: U.S. Department of Commerce, various issues); and data supplied by various steel firms and equipment suppliers. See Appendix C.

mates for existing and state-of-the-art facilities. This will allow us to assess the viability of the greenfield route for eliminating the cost gap that U.S. integrated producers now face. Table 7–5 compares total costs for mini-mills and integrated plants on a current and state-of-the-art basis, in 1981 dollars. The plants described in this table are assumed to attain the minimum optimal scale; potential cost savings are calculated considering all major input categories, not just labor requirements.

As the data presented in Table 7–5 show, the move to state-of-the-art operations could reduce real production costs in mini-mills; higher capital costs are offset by other savings. The same cannot be said for integrated operations, however. In state-of-the-art integrated plants producing typical products, such as cold-rolled sheet, significant savings in operating costs are more than offset by increased capital costs. For these products, real costs rise in state-of-the-art integrated plants compared to average costs in existing plants. Proponents of the greenfield strategy could argue that depreciating capital costs over a longer time period could weaken the negative effects of high capital costs sufficiently to allow new integrated plants to enjoy a real cost advantage over existing mills. We would argue, however, that the tempo of technical change is such that most facilities should be substantially rebuilt after a fifteen-year life. Furthermore, these tables do not include any return on equity, which is assumed to provide 50 percent of the funding for construction. A satisfactory return on equity would boost new plant costs even further and would exacerbate the difference between integrated and mini-mill plants since integrated plants are more expensive to build. No reasonable set of assumptions can be selected to generate an attractive return on investment for new integrated plants.

State-of-the-art integrated operations do realize real cost savings in mini-mill products such as wire rods, but these savings do not lessen the cost advantages that mini-mills enjoy. A new integrated plant producing wire rods would have costs 30 percent above the present mini-mill average and 39 percent above the costs attainable in state-of-the-art mini-mills.

This analysis leads to two disturbing conclusions for U.S. integrated producers. The first concerns the competitive standing of integrated producers vis-a-vis their mini-mill competitors. In products for which mini-mill techniques are appropriate, they have a significant and inherent advantage in labor productivity. Even with state-of-the-art techniques, labor requirements in integrated facilities are roughly twice those of mini-mill operations. Hence, the mini-mill advantage is not based on lower wage rates, newer facilities, entrepeneurship, and so on—although all of these factors represent real advantages that further strengthen the superior competitiveness of mini-mill operations. Rather, the mini-mill route is simply a more efficient technical path to the production of certain products. Firms that attempt to compete against mini-mills by using even the most efficient possible integrated techniques will still face a seemingly insurmountable disadvantage. Given the fact that state-of-the-art operations are almost certainly beyond the reach of most U.S. integrated plants, the attempt to retain market share in mini-mill products is hopeless. Integrated firms that are committed to future production of wire rods, bars, light structurals, and so on must adopt mini-mill techniques for those products. The greenfield strategy offers no protection for integrated producers here.

The second and perhaps more disturbing conclusion involves the competitive standing of U.S. integrated producers versus their foreign competitors, especially the Japanese. Massive modernization in its purest form—the construction of greenfield plants of the minimum efficient scale—would not improve their competitiveness against their principal foreign competitors. Instead, it would increase the competitive disadvantage now facing domestic integrated steelmakers in typical integrated products. The problem is that new integrated plants are extremely expensive, costing well over $1,000 per annual ton of capacity. A hypothetical plant of the minimum efficient scale (4 million tons capacity) would thus cost over $4 billion to build. The financing costs involved in expenditures of this magnitude swamp the efficiency benefits provided by state-of-the-art operations.[6]

If anything, inflation in capital costs and high interest rates are increasing the disadvantages of the massive-modernization strategy and, concomitantly, any attempt to approximate greenfield construction at existing sites. The burden of high capital costs is not likely to abate; the more difficult it becomes to raise funds, the less viable is the greenfield route to improved performance. Integrated producers in the United States have grown accustomed to bemoaning the lack of funds that prohibits them from building greenfield plants or raising the efficiency of existing facilities to state-of-the-art levels. This misses the point. The problem is not the availability of funds but the fact that the industry's traditional investment strategy is no longer economically viable. Funds are available for investments that provide sufficient cost advantages, as is shown clearly by the expansion of the mini-mill sector. Unsurprisingly, financial markets will not support an investment strategy that results in higher costs.

Insofar as the massive-modernization strategy draws support from the record of the Japanese steel industry, the recent performance of that industry should be enough to provoke some doubts. Indeed, the historical record since 1975 suggests that Japanese steel producers are in the throes of a major structural crisis. The problems of the world steel industry since 1975 indicate that conditions have changed drastically and that the halcyon growth of the postwar period has ended. The ability to adapt to these altered circumstances will be just as crucial for Japanese producers as for their American competitors; Japan's present preeminence, somewhat similar to the status enjoyed by the U.S. steel industry in the 1950s, may prove to be a burden in the future. It would be tragic if the U.S. steel industry were to idealize the performance of an industry that has passed its peak.

All of these considerations suggest that the steel industry must develop a strategy more appropriate to the conditions it faces. Long-standing prejudices about the requirements for improved performance must be reevaluated. Investments must target only those factors that are essential, since capital constraints will ensure that the penalties for mistaken investment decisions are steep. Unfortunately, the commitment to monumentalism is so engrained that very little effort has been devoted to defining or assessing alternative strategies.

IMMEDIATE DETERMINANTS REVISITED: SCALE VERSUS TECHNOLOGY

In order for investments to have the greatest possible impact, they must be selected according to an evaluation of the causal hierarchy among the immediate determinants of performance. We will begin analyzing this causal structure by reexamining the productivity data presented in Chapters 5 and 6. This is an exercise in comparative statics. We will then turn directly to the question of scale economies, concentrating on both labor and capital requirements for different plant sizes. These estimates can then be used to compare the relative benefits of scale and capacity utilization. Finally, using data provided by steel firms and equipment suppliers, it is possible to estimate economies of scale for current state-of-the-art plants and also for state-of-the-art plants as they would have been constructed in the early 1960s. These estimates provide the basis for a dynamic comparison between technology and scale in terms of their consequences for performance, and this theme will be addressed after our discussion of economies of scale per se.

A Comparative Static Assessment

In Chapter 6, economic choices, technology, and scale were described as the principal immediate determinants of operating performance. Unfortunately, the complex interrelationships among these factors makes it difficult to isolate and define the relative contribution of each to overall productivity growth. Such a project is further complicated by the practical impossibility of strictly quantifying technical change. Rough proxies such as the capital-labor ratio, patent filings, and so on are generally used for this purpose, but these proxies inevitably fail to capture important aspects of the process they are designed to measure. These considerations make it difficult to allocate productivity improvements among the three categories of direct causes that we have discussed. Nevertheless, the data presented in previous chapters, when informed with a qualitative understanding of the actual processes involved, can be interpreted to draw valid inferences about the relative importance of the factors which directly determine productivity growth.

Integrated Techniques: Japan versus the United States. Table 7–6 presents a static assessment of the direct causes of the Japanese productivity advantage over the United States in the integrated production of cold-rolled sheet and wire rod. This table allocates the 1980 productivity differences between the U.S. and Japanese steel industries among the three categories of scale, technology, and economic choices. As in Chapter 6, technological differences are broken into two categories: major technical changes and operating techniques. Improvements that encompass several factors have been allocated among each of the relevant categories or have been assigned according to the authors' assessment of which aspect is primary. Scale is a residual category; it captures some differences that are probably due to technological factors.

If we turn first to cold-rolled sheet, we find that the total Japanese advantage in 1980 amounted to 1.36 MHPT, even though the United States enjoyed an advantage of 0.85 MHPT due to economic choices made by the two industries. As was argued in Chapter 6, economic choices reflect the inherent comparative advantages and disadvantages of an industry. For Japan, this implies a reliance on foreign sources of raw materials and on foreign sales. As a result, Japanese steel producers use less scrap than do U.S. firms, and their labor requirements for handling and shipping are greater. Such economic choices generally represent a disadvantage for Japanese producers in spite of their greater efficiency in each process.[7]

Technological differences are the predominant cause of superior Japanese productivity, providing 70 percent of the net advantage enjoyed by the Japanese industry in 1980. Better operating techniques contribute slightly more to this result than do differences in the adoption of major technological advances. More extensive use of the basic oxygen furnace (BOF) and continuous casting—including better casting performance and casting-related yield improvements—are the two major technological dif-

Table 7–6. Definable Causes of Japanese Productivity Advantages over the U.S. Integrated, 1980.

	Cold-Rolled Sheet		*Wire Rod*	
	MHPT	*Share (%)*	*MHPT*	*Share (%)*
Economic choices				
Input use	−0.45		−0.44	
Package & overhead	−0.40		−0.27	
Subtotal	−0.85	−62	−0.71	−32
Technological differences				
A. Major new technologies				
Steelmaking: more BOFs	0.18		0.17	
Continuous casting				
Better performance	0.23		0.24	
More CC	0.20		0.44	
Yields: CC related	0.11		0.09	
Subtotal	0.72	53	0.94	43
B. Operating techniques				
Material handling	0.04		0.04	
Coke rates	0.06		0.06	
BF: coke rate related	0.07		0.07	
Ingot practices	0.01		0.01	
Yields: Not CC related	0.26		0.20	
Primary rolling	0.04		0.80	
Cold finishing	0.22		–	
Package & overhead	0.20		0.20	
Subtotal	0.82	60	1.38	62
Scale differences				
Coke ovens	0.16		0.16	
Sinter	0.08		0.08	
Blast furnaces	0.21		0.19	
BOF	0.07		0.07	
Hot rolling	0.15		0.10	
Subtotal	0.67	49	0.60	27
Total	1.36	100	2.21	100

ferences, and these have been absolutely crucial to Japanese success. Nevertheless, one cannot attribute the Japanese performance to these differences alone. Table 7–6 shows that there is a wealth of more incremental differences (operating techniques) that have significantly increased the Japanese lead in productivity. These include the following advantages: better material handling, improved coke rates, generally better yields, improved cold finishing, and more efficient packaging, shipping, and administration.

Scale is also a significant contributor to the Japanese productivity advantage in cold-rolled sheet, although its effects are less dramatic than those due to economic choices or technology. Technological differences, broadly defined, are more than twice as significant as scale, and the disadvantages connected with economic choices more than offset the advantages associated with the greater scale of the Japanese steel industry. In general, these conclusions are also applicable to integrated production of wire rods; the importance of technological differences,

Table 7–7. Definable Causes of U.S. Mini-mill Productivity Advantages over Integrated: Wire Rod, 1980.

	MHPT	*Share (%)*
Economic choices		
Input use	1.27	43
Technological differences		
A. Major new technologies		
Steelmaking	–	
Continuous casting		
Better performance	0.23	
More CC	0.63	
Yields: CC Related	0.15	
Subtotal	1.01	35
B. Operating techniques		
Yields: not CC related	0.04	
Primary rolling	0.65	
Package & overhead	0.21	
Subtotal	0.90	31
Scale differences		
Steelmaking	−0.20	
Rod mill	−0.05	
Subtotal	−0.25	−9
Total	2.93	100

especially in operating techniques, is even more pronounced for that product.[8]

The United States: Integrated Versus Mini-mill. Table 7–7 describes the causes of the productivity differences between U.S. integrated firms and mini-mills in the production of wire rods. In 1980, the mini-mill advantage totaled 2.93 MHPT (45%). The lack of primary processes–an economic choice–provided 43 percent of the net mini-mill advantage. While this difference is highly significant, the principal cause of the mini-mill advantage lies in the area of technology, especially more and better continuous casting. Differences in operating techniques (better yields, superior primary rolling, and tighter control of overhead) are almost equally important. These reflect characteristics of mini-mill operations that have already been discussed: simpler processes, the elimination of redundancies and crutches, restricted product lines, and greater attention to cost competitiveness. The mini-mill technical advantage also stems from newer facilities, as was the case for Japanese integrated operations.

Perhaps the most significant result in regard to the mini-mill advantage is the surprisingly insignificant role played by scale. This in part reflects the fact that integrated wire rod mills are little larger than those in mini-mills. Nevertheless, integrated *plants* are much larger, and one might suppose that their wire rod productivity would benefit from this. Yet the integrated economies of scale in rod products stem from primary processes–an advantage that is nullified by mini-mills' reliance on the electric furnace. The only significant scale advantage for integrated producers is in steelmaking, but the advantage of the BOF over the electric furnace as a melting vehicle has narrowed considerably in recent years as a result of numerous technical improvements. In toto, scale amounts to an insignificant 9 percent disadvantage for mini-mills.

Comparative Static Results. While one could quibble with individual entries in these tables, they clearly demonstrate that technical change is the predominant direct engine of productivity growth. They also suggest that technical differences, broadly defined, are even more predominant in a mini-mill product like wire rod than in an integrated product like cold-rolled sheet. This is true whether the comparison is undertaken between two integrated industries (U.S. vs. Japan) or between an integrated and a mini-mill sector (in the U.S.). This suggests that technical change has been somewhat more rapid in the set of processes that are used to produce mini-mill products.

The technological foundation of superior Japanese performance is somewhat surprising, since the Japanese advantage over U.S. producers is commonly linked to the enormous scale that is characteristic of Japanese steel plants. Technological differences are harder to quantify and less immediately apparent. Yet they far outweigh the effects of scale per se. Scale may appear to be more important than it truly is because, as we have argued, many technological improvements manifest themselves as changes in scale. Scale and size are different concepts. In regard to the mini-mill advantage over integrated producers, the extent to which technology overshadows scale is less surprising, although the minor benefits of the integrated scale advantage are striking.

The technological advantages of Japanese firms and mini-mills over U.S. integrated producers cannot be ascribed solely to major, and expensive, differences in facilities. Within the category of technical change, these tables suggest that the cumulative effect of myriad incremental technical improvements, categorized as operating techniques, actually seem to be more important than major technical advances—although there is obviously some synergism between these two categories. Especially in the Japanese case, many of these minor technical advantages are made possible by the use of data processing (for process control, product control, energy savings, etc.) and by the continuous monitoring of performance—as is evidenced by the very availability of the detailed labor data which were presented in Chapter 5.

Differences in operating techniques also reflect differences in attitudes and behavior. U.S. integrated firms are less attuned to the early recognition and rapid implementation of technical improvements—a point shown most clearly by comparing continuous casting shares. Furthermore, constant improvements in operating practices are a more conscious goal of both management and labor in Japan and in American mini-mills. It is true that the Japanese, at least, have spent more than U.S. integrated firms for modernizing facilities. Nevertheless, it is doubtful that this would have generated the observed record of productivity growth had it not been for a strong subjective commitment to constantly raising the level of performance.

Unfortunately, the overriding importance of technical change that has been described by these static comparisons disappears if one attempts to construct a similar table referring to changes over time. If one allocates the productivity differences in Japan between 1958 and 1980, for instance, scale appears to be the overwhelmingly predominant direct source of improvement. This is due to the fact that technological advances manifest themselves as increased throughput (i.e., greater scale) over time. The merging of these two categories reflects the fact that both scale and technology, unlike economic choices, are *dynamic* determinants of productivity growth, with complex and synergistic benefits.

A partial indication of the dynamic contributions of these categories can be developed by comparing present performance with new-plant potential. This is done in Table 7–8, which presents such a comparison for integrated production of cold-rolled sheet and mini-mill production of

Table 7–8. Definable Causes of New-Plant Advantages over Current U.S. Plants.

	Integrated Cold-rolled Sheet		*Mini-mill Wire Rod*	
	MHPT	*Share (%)*	*MHPT*	*Share (%)*
Economic choices				
Input use	0.03	1	–	
Technological differences				
A. Major new technologies				
Steelmaking (more BOFs)	0.12		–	
Continuous casting				
Better performance	0.27		0.16	
More CC	0.39		0.29	
Yields: CC related	0.21		0.04	
Subtotal	0.99	34	0.49	32
B. Operating techniques				
Coke rates	0.07		–	
BF: coke rate related	0.07		–	
Yields: not CC related	0.22		0.01	
Cold finishing	0.38		–	
Package & overhead	0.43		0.27	
Subtotal	1.17	40	0.28	19
Scale Differences				
Coke ovens	0.16		–	
Sinter	0.06		–	
Blast furnaces	0.21		–	
Steelmaking	0.08		0.27	
Hot rolling	0.20		0.47	
Subtotal	0.71	25	0.74	49
Total	2.90	100	1.51	100

wire rod. This table confirms the general conclusions presented above, although it was assumed that economic choices vary little, so that this category ceases to have a significant effect. Technological differences predominate in both cases, although scale is almost as important for mini-mills. This is not surprising, given the extent to which mini-mills fall short of the minimum efficient plant scale—400,000 tons vs. 750,000 tons. Integrated plants, on the other hand, have an average scale that already matches the minimum optimal level, although this masks significant scale deficiencies in regard to specific facilities, especially coke ovens, blast furnaces, and hot strip mills.

Table 7–8 illustrates the importance of more and better continuous casting (plus the associated higher yields), especially for integrated products. Operating techniques are also important, although more so in integrated plants than in mini-mills. This again suggests that mini-mills are much closer to state-of-the-art performance and that the simpler configuration of mini-mills makes it easier to achieve performance improvements.

Economies of Scale

The possibility that scale plays a secondary role in determining performance conflicts with the traditional view of the steel industry as one characterized by significant economies of scale. Although this is still the case for integrated techniques, it is much less true for mini-mill operations. Even for integrated plants, the minimum optimal scale differs with different product mixes, and there is little evidence (although much presumption) that the mammoth size of some of the newer Japanese plants—up to 14 million tons annual capacity—is necessary for optimal performance. According to our analysis, the efficiency of such plants is due much more to their new technology than to their large scale.

The Structure of Scale Economies: Capital Costs and Labor Requirements. Further evidence on the role played by scale can be developed by directly assessing the structure of scale economies in contemporary plants. In Chapter 6 we presented data on the actual historical relationship between scale and productivity for various facilities in the U.S. and Japanese steel industries. Those data measured efficiencies of scale in labor requirements; these must be supplemented with estimates of the effects of scale on capital costs in order to assess the overall importance of scale at the facility level. These facility estimates can then be combined to describe the plantwide impact of scale increases and to evaluate the effects of different product mixes on scale economies. Only when all of these elements have been combined is it possible to draw useful conclusions about how scale can be adjusted to foster the best possible performance.

In the first part of this chapter, engineering estimates were used to describe the total costs of a state-of-the-art plant. Unlike the time series data presented in Chapter 6, these estimates compared the performance of different sizes of plants at the same point in time. Similar estimates have been collected relating to unit capital costs for different sizes of

Figure 7–1. Economies of Scale in New Plants: Integrated.

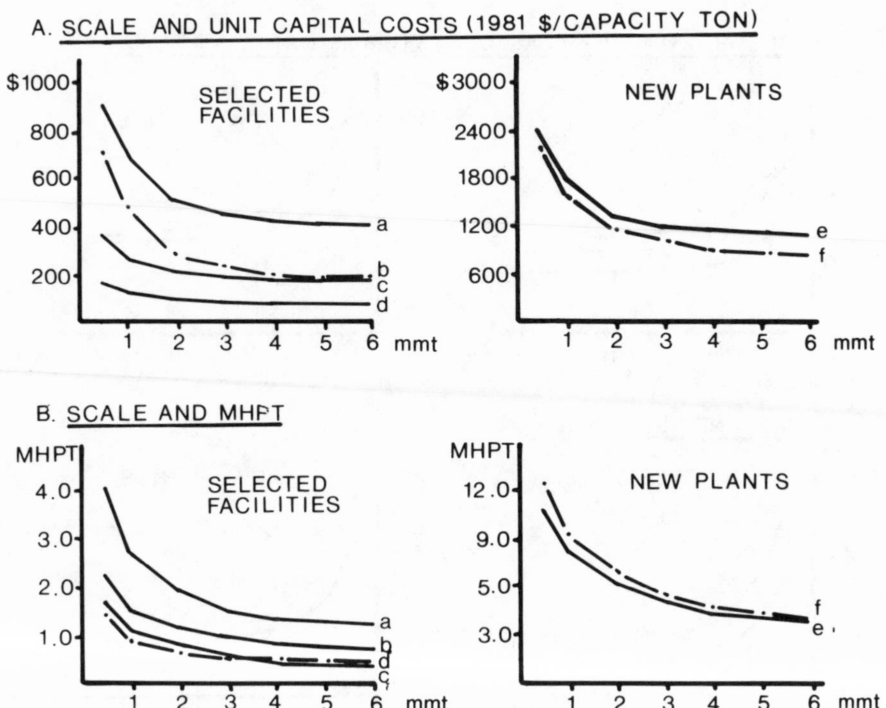

a. Coke to slab.
b. Hot strip mill.
c. Blast furnaces.
d. BOF (melt).
e. Cold-rolled sheet.
f. Cold-rolled sheet, Plate, Bar.

Sources: See Table 7–1.

facilities and plants. These are shown in Figure 7–1, which refers to integrated plants and which also describes the relationship between scale and labor requirements, derived mainly from the data base that was analyzed in previous chapters. Based on the data shown in Figure 7–1, we estimate that the minimum optimal scale for an integrated plant producing cold-rolled sheet is four million tons annual capacity, although most scale benefits are attained at three million tons. This is largely determined by scale economies in hot strip mills, where the minimum optimal scale is about four million tons, rather than in the hot end

Figure 7–2. Economies of Scale in New Plants: Mini-mill.

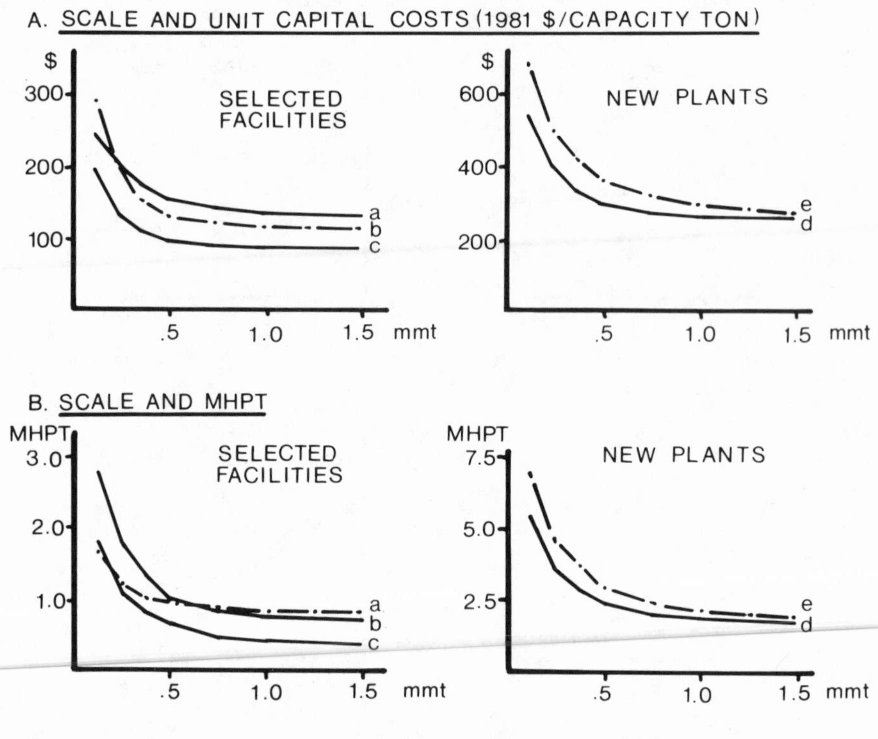

a. Rod mill.
b. Electric Furnace and Billet Caster.
c. Electric furnace.
d. Wire Rod.
e. Wire Rod and Bar.

Sources: See Table 7–1.

(ironmaking and steelmaking), where the minimum optimal capacity at the plant level is three million tons or less. At the hot end, the minimum optimal scale of an ironmaking or steelmaking shop is at least twice the level required for an individual furnace. This is because furnaces invariably require periodic maintenance work (e.g., for relining) which can last up to several months. Furthermore, as the production process becomes more continuous, especially due to continuous casting, the batch pattern of production by a single furnace can disturb the continuity of other operations. For these reasons, multiple furnace shops are a virtual necessity.

A four-million ton integrated plant producing cold-rolled sheet would realize about 90 percent of the labor productivity and unit capital cost benefits achieved in a ten-million ton plant, assuming that both plants are new and both are operating at the same rate. This slight disadvantage would probably be more than offset by the fact that a smaller facility would be more adaptable and in closer contact with its markets, so that it would therefore be less vulnerable to fluctuations in demand. As a result, operating rates would tend to be higher in the smaller plant, with salutary consequences for performance.

Figure 7–2 provides engineering estimates of unit labor requirements and unit capital costs for variously sized mini-mills. As these graphs show, the minimum optimal scale for a mini-mill producing wire rods is approximately 750,000 tons, although most of the scale benefits are realized at 500,000 tons.[9] None of the mini-mill facilities, including the rod mill itself, appear to enjoy substantial scale efficiencies beyond this output level. A 750,000-ton plant would be small enough to adapt to market forces and to maintain relatively high operating rates. This greater adaptability represents a real, albeit unquantifiable, advantage for the smaller plant.

Economies of scale are altered by the choice of output. According to the data presented in Figures 7–1 and 7–2, the minimum optimal scale is quite large for most integrated facilities and for integrated plants relative to mini-mill standards. However, the relationship between scale and efficiency in blast furnaces, BOFs, and rod mills imply that the economies of scale for integrated production of barlike products (such as wire rods) are significantly below those that characterize flat-rolled products. For the production of barlike products, the minimum optimal scale is defined by scale economies at the hot end; these suggest a minimum optimal plant size of three million tons.

More significantly, perhaps, economies of scale are also altered by the variety of the product mix. This too is illustrated in Figures 7–1 and 7–2. For both mini-mills and integrated plants, the adoption of a broader product mix increases the minimum optimal scale. An integrated plant

Figure 7–3. Comparative Cost Effects: Shifts in Scale versus Shifts in Capacity Utilization[a]

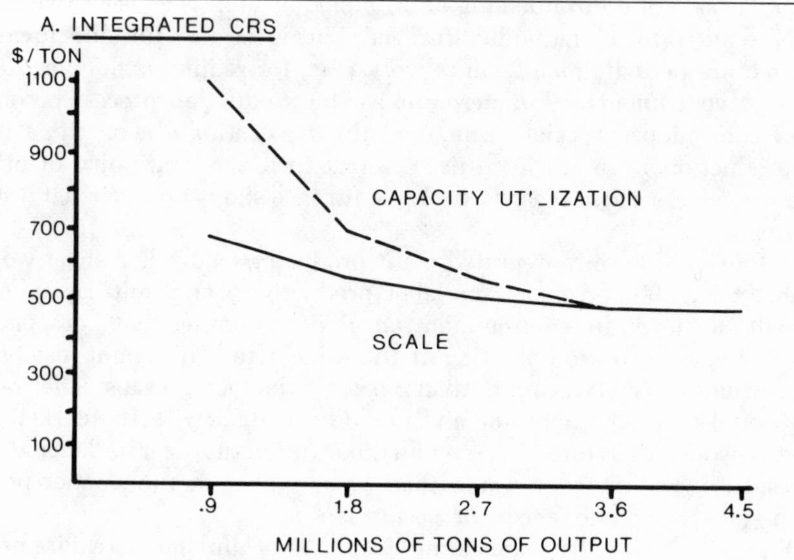

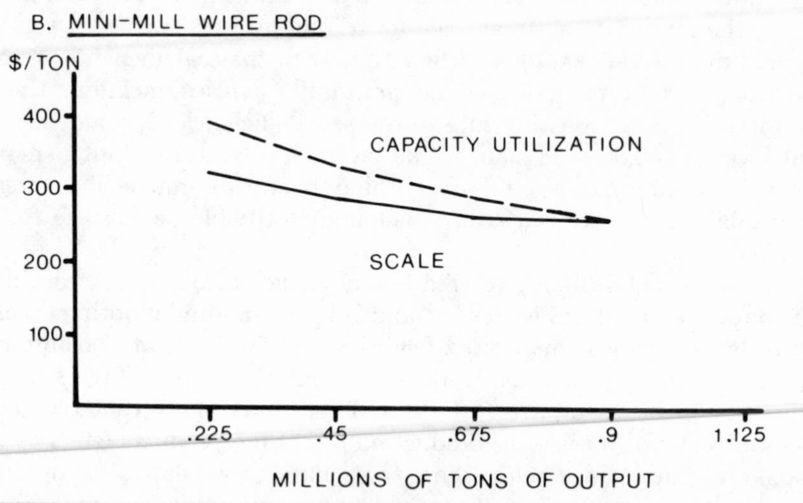

a. Each diagram shows total unit production cost at alternative output levels, either with different plant sizes (at 90% CU) or with given plant size (at alternative rates of capacity utilization). The former is referred to as "scale," the latter as "capacity utilization." The plant sizes used for evaluting the effects of shifts in capacity utilization are 4 million tons for integrated production of cold-rolled sheet (CRS) and 1 million tons for mini-mill production of wire rods (WR).

Sources: See note 3 of this chapter and Appendix C.

producing three diverse products must have an annual capacity greater than the four million tons required in a single-product plant in order to attain 90 percent of the potential economies of scale. Similarly, the minimum optimal scale for a mini-mill producing two diverse products exceeds the 750,000 tons required in a one-product plant. This confirms the view that specialization is an alternative to increased capacity as a means of achieving the benefits of scale economies.

The static analysis of the relative importance of scale, technology, and economic choices suggested that technology is the most crucial determinant of performance. The inference that scale is less significant is further strengthened by the identification of minimum efficient scale at levels far below the preferred capacity of greenfield Japanese plants. Moreover, it is possible to reduce the minimum optimal scale even further by specializing in a limited range of products. While a smaller plant sacrifices some marginal scale-related efficiencies, these may be outweighed by the advantages of greater market flexibility.

Scale and Capacity Utilization. Smaller scale plants are likely to enjoy lower transportation costs and to possess more clearly defined markets, and this suggests that there is a negative correlation between scale and operating rates. Insofar as a smaller plant is able to achieve higher operating rates, the scale advantage of the larger plant will be quickly eliminated. This is shown in Figure 7–3, which compares the relative effects of scale and capacity utilization for integrated production of cold-rolled sheet and for mini-mill production of wire rods.

The scale curve in Figure 7–3 measures true economies of scale; it describes total costs for several different plant sizes operating at a 90 percent rate. The capacity utilization (CU) curve in Figure 7–3 measures total costs at a new, optimally sized plant (a one-million ton mini-mill and a four-million ton integrated plant) at various operating rates (along the capacity utilization curve). These curves make it possible to identify the sensitivity of operating costs to shifts in scale and to shifts in operating rate. For integrated plants producing cold-rolled sheet, total unit costs of production decline 28.5 percent if plant size increases from one to four million tons capacity. For mini-mills producing wire rods, total unit costs decline 18.5 percent if plant size quadruples from 250,000 to 1 million tons. The greater sensitivity of integrated plants to scale changes stems from the fact that labor and capital costs, which are the only cost components significantly affected by scale differences, comprise 51 percent of new plant costs for integrated production of cold-rolled sheet. In new mini-mills producing wire rods, labor and capital costs make up only 20 percent of total unit costs. The effect of scale

Table 7–9. Scale versus Technology and Labor and Capital Savings in New Plants[a] (90% operating rate).

	Scale	1960s Technology			1980s Technology		
	(million tons)	*Capital Costs ($/t)*	*Labor Use (MHPT)*	*C/L ($/MH)*	*Capital Costs ($/t)*	*Labor Use (MHPT)*	*C/L ($/MH)*
Integrated (CRS)[b]	2.00	1715	8.90	193	1475	6.00	246
	3.00	1541	7.35	210	1276	4.85	263
	3.30	1522	6.95	219	1250	4.63	270
	4.00	1472	6.50	226	1205	4.30	280
	5.00	1386	6.10	227	1147	4.05	283
Mini-mill (WR)[c]	0.25	890	10.45	85	450	3.50	129
	0.40	711	8.35	85	372	2.81	132
	0.50	651	7.40	88	327	2.40	136
	0.75	582	6.15	95	300	2.00	150
	1.00	545	5.70	96	286	1.90	151
Integrated (WR) A. [c]	0.50	1549	12.90	120	1030	6.40	161
	0.75	1321	10.30	128	880	5.15	171
	1.00	1201	9.00	133	800	4.50	178
B. [d]	2.00	1015	8.35	122	675	3.95	171
	3.00	965	7.80	124	640	3.70	173
	4.00	945	7.45	127	625	3.55	176

a. Technology includes operating practices.
b. Plant size, all hot- or cold-rolled sheet.
c. Plant size, all wire rod.
d. Plant size, mix of rod, bar and light structural, 0.75 million tons wire rod.
Sources: As in Figure 7–4.

changes on labor and capital costs, rather than total costs, is proportionately the same in both types of plants.

To illustrate the importance of capacity utilization on optimally sized plants, unit costs were estimated for a one-million ton mini-mill and a four-million ton integrated plant operating at output levels which correspond to the various plant sizes referred to by the scale curves in Figure 7–3. As these diagrams show, shifts in operating rate have a greater impact on costs than do changes in scale. In a four-million ton integrated plant, for instance, total unit costs fall 65 percent if operating rate and output quadruple from one to four million tons, while a corresponding quadrupling of plant size (assuming a 90% operating rate) reduces unit costs by 28.5 percent. In the mini-mill, a fourfold increase in operating rate, boosting output from 250,000 tons to 1 million tons, lowers unit costs by 31 percent, while a corresponding increase in scale, again assuming a 90 percent operating rate for each plant size, provides only an 18.5 percent saving.

High rates of capacity utilization thus appear to be more important than scale economies in terms of overall performance. The effects of both these variables are relatively more significant in integrated plants than in mini-mills, but the dominance of capacity utilization over scale is also relatively greater for integrated techniques. The ratio of cost savings from higher operating rates versus cost savings from greater scale is 2.25 in the integrated sector and 1.68 in the mini-mill sector (for a comparable quadrupling of output, as in Figure 7–3). This confirms the necessity of striving for higher rates of capacity utilization as a long-run goal—especially in integrated plants.

Scale economies are important, but not at the expense of high operating rates. Firms must seek to attain scale economies that are geared to the market, a goal that may be achieved via specialization, which does not conflict with high operating rates, rather than via plant size. By avoiding excessive product overlap among competing plants, both higher scale and superior operating rates may be realized in the long run.

Scale versus Technology: a Dynamic Comparison

Engineering estimates of labor requirements and capital costs in new mills of the early 1960s and in new mills of the early 1980s can be used to isolate the relative contributions of scale and technical change to improved performance. Presumably, performance improvements attained by moving from a 1960s plant to a 1980s plant of the same scale can be ascribed to the effects of technological change alone. Alternatively, the effects of scale changes can be evaluated while holding technology fixed.

Figure 7–4. The Impact of Technical Changes on Capital Costs and Labor Use (90% operating rate; 1981 $/ton of capacity).

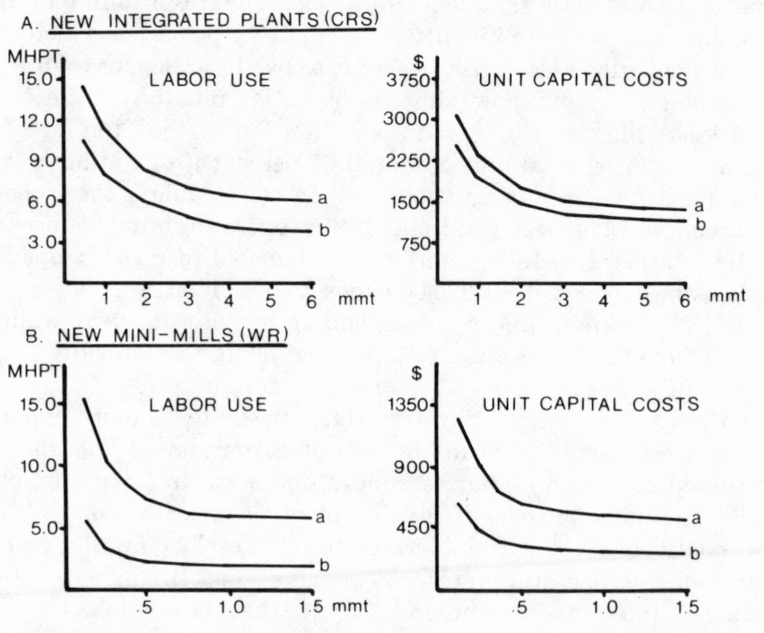

a. 1960's technology.
b. 1980's technology.
Sources: See Table 7–1.

This is one means of disentangling the effects of scale and technology. The data developed to make such a comparison are presented in Figure 7–4 and in Table 7–9; these illustrations describe engineering estimates of scale economies (in 1981 dollars and MHPT) for new plants using 1960s technology and using 1980s technology.

Essentially, we are comparing the costs and performance of a plant as it would have been designed circa 1961 with the preferred plant designed in 1981. Since our purpose is to isolate the effects of technical change on productivity and on unit capital costs and since continuous casting is the most crucial technical change that occurred during this period, we have assumed that the earlier plant predates the commercialization of continuous casting. Billet casters could conceivably have been installed in mini-mills during the early 1960s, but there is little doubt that an integrated plant designed in the early 1960s would not have included a continuous caster. Bethlehem's Burns Harbor plant, on which construction was begun in 1962, did not have a continuous caster until 1975. For the 1980s plant, we assume 100 percent continuous casting.

Differences in productivity performance and in unit capital costs from the early 1960s to the early 1980s are described in Figure 7–4 and Table 7–9. Assuming a constant operating rate of 90 percent, these illustrations show estimated capital costs and labor requirements at various scales for the production of cold-rolled sheet and wire rods—the latter using both integrated and mini-mill techniques. On the basis of these estimates, a series of capital-labor ratios have also been constructed and included in Table 7–9. While these data are estimates rather than observed results, they are adequate to support the conclusions that we will draw.

In general, the results for 1980s plants versus 1960s plants indicate that technical changes have substantially reduced both real unit capital costs and labor requirements. This has been true for both integrated and mini-mill plants, although the gains have been far more significant in the latter case. Furthermore, in both integrated plants and mini-mills, the reduction in labor requirements exceeded the capital cost savings.

For a more detailed analysis, let us turn first to labor productivity. We have already identified four million tons as the minimal optimal scale for integrated operations producing flat-rolled products (e.g., cold-rolled sheet) and 750,000 tons as the minimum optimal scale for mini-mill operations producing typical mini-mill products (e.g., wire rods). At the minimum optimal scales, Table 7–9 shows that the transition from 1960s technology to 1980s technology offered a reduction in labor requirements of over 65 percent for mini-mills (wire rods) and roughly 33 percent for integrated plants (cold-rolled sheet). The productivity benefits of technological progress described in this table appear to far outweigh those provided by scale increases. Average mini-mill capacity in 1964 amounted to 250,000 tons. By 1980, this had increased to 400,000 tons, providing a productivity gain of 20 percent. Boosting this to the minimum optimal level would improve productivity by about another 28 percent using both 1960s and 1980s technology. For integrated producers, an increase from the average plant size in 1964 (3.3 million tons) to the minimum optimal scale of four million tons—which is, incidentally, the average 1980 plant size—reduces labor requirements by 7.5 percent using both 1960s and 1980s techniques.

The fact that the average scale of U.S. integrated facilities is so close to the minimum optimal scale is misleading, however. It does not imply that corresponding scale efficiencies are attained, for while the steelmaking capacity of U.S. integrated plants is close to the minimum optimal level, finishing facilities rarely approach this standard (see Chapter 6). The hypothetical plant used in this analysis is assumed to have one hot strip mill, with an annual capacity of four million tons. The actual average size of U.S. hot strip mills is only half that figure (see Table 3–4). This is an indication that U.S. integrated plants are insufficiently specialized to realize optimal scale efficiencies; concentration on a more limited product range would make it possible to gain adequate economies of scale in plants in which steelmaking capacity does not exceed the present average. In terms of our analysis, one consequence of the lack of specialization in these plants is that the effective scale of actual integrated plants in the United States has been somewhat overstated in the preceding paragraph, although this caveat does not affect the data presented in Table 7–9.

Nevertheless, it is still evident that for both integrated plants and mini-mills, the benefits of technological progress far exceed those generated by scale increases. This confirms our earlier assessment. Technological change has been the most significant source of improved performance in the steel industry, in spite of the common view that the industry is technologically stagnant and therefore that scale effects predominate. Moreover, the comparisons presented in Table 7–9 show clearly that technical progress, as related to productivity improvements, has been most robust in the mini-mill sector. As we have shown, actual productivity improvement has been more pronounced in the mini-mill sector than in the integrated sector; Table 7–9 confirms that this would have been the case even if both sectors had been able to modernize at the same rate.

Similar results characterize the potential trend in capital costs. At the minimum optimal scale, unit capital costs in mini-mills have been, or would be, reduced by almost 50 percent by the transition from the technology of the 1960s (prior to continuous casting) to that of the 1980s. For integrated production of cold-rolled sheet, the reduction amounts to about 20 percent. Scale increases from the average 1964 capacity to the 1980 level (the minimum optimal) have reduced real unit capital costs by less than 5 percent for integrated producers. In mini-mills, the increase in scale from 1964 to 1980 levels has reduced unit capital costs by 17 percent. A further increase to the minimum optimal scale would have reduced unit capital costs by another 18 to 19 percent.

Thus, for both integrated and mini-mill producers, for both labor requirements and capital costs, technological change appears to be the most crucial determinant of performance. Because of the close link between technology and scale–defined as throughput capacity–increased competitiveness is often associated with greater scale as well as with newer techniques; but it is the latter that are primary. As a result, the opportunity to implement new technologies must be preferred over the opportunity to increase size, especially when funds are limited.

THE MINI-MILL ADVANTAGE

The data presented in Table 7–9 provide us with a deeper understanding of why mini-mills have had such success against their integrated competitors. In fact, the mini-mill sector, rather than the Japanese steel industry, offers the best model for revitalizing the U.S. steel industry as a whole. Mini-mills have prospered, sometimes spectacularly, under just those conditions that have sapped the strength of integrated firms. Lower costs, more modern facilities, and similar characteristics are associated with the mini-mill sector, but these are results rather than causes. The underlying advantage of this sector is the fact that it has enjoyed more rapid and more fundamental technical changes than has its integrated counterpart. The problem for integrated producers vis-a-vis the mini-mills is not that the mini-mills are closer to the technological frontier, although this does represent a significant static disadvantage for integrated firms; rather, the more threatening difference is that the frontier is moving more rapidly for the mini-mill sector as a whole. Mini-mills now occupy the technological "fast lane" for steel production.

The more rapid rate of technical change in the mini-mill sector is linked to the fact that mini-mills have adopted or generated a specific type of technical change. This can be illustrated by interpreting the capital-labor ratios presented in Table 7–9. As was discussed in Chapter 6, increases in the capital-labor ratio reflect capital-deepening investment, which is generally a proxy for the adoption of significant technological improvements. Table 7–9 describes capital-labor ratios for three operations: integrated production of cold-rolled sheet, integrated production of wire rods, and mini-mill production of wire rods. In each case, technological progress (moving from 1960s to 1980s technology, holding scale constant) has the expected effect of increasing the capital-labor ratio. Somewhat surprisingly, increases in scale also raise the capital-labor ratio. Greater scale is generally thought to be relatively capital saving, but the data presented in Table 7–9 suggest that increased scale tends to be relatively labor saving up to the minimum optimal scale, after which further increases in scale have a neutral effect on the capital-labor ratio.[10]

If we compare the capital-labor ratios for 1980s and 1960s technology while holding scale constant, Table 7–9 shows that the most dramatic increase in this ratio occurred in the mini-mill sector, followed by integrated production of wire rods. The increase in the capital-labor ratio for the integrated production of cold-rolled sheet has been relatively small. The potential substitution of capital for labor, defined by the available state of the art, has been greatest in the mini-mill sector. At the same time, however, mini-mill techniques offer the greatest prospect for reducing unit capital costs. Mini-mills thus enjoy the best of both worlds: they have been able to employ what are relatively and absolutely the most labor-saving techniques, yet they have also been able to attain the greatest absolute savings in unit capital costs. This felicitous combination fuels the engine of mini-mill expansion. The mini-mill sector has been created by—or has created—a pattern of technological innovation that has dramatic effects on several fronts. Mini-mills have solved the dilemma facing integrated producers—namely, how to generate significant performance improvements in the face of increasingly severe capital constraints—by adopting technologies that provide dramatic savings in both labor and capital. *The development and rapid implementation of significant capital-saving technologies is the foundation of the mini-mill advantage.*[11]

Consciously or unconsciously, all of the characteristic features of mini-mills serve to accentuate the impact of capital-saving technical changes. The simplified and streamlined structure of processes that characterizes mini-mills ensures that technical breakthroughs in one process (melting, casting, or rolling) will have a significant effect on overall performance. This feature has boosted the impact of continuous casting for mini-mills relative to their integrated counterparts. In terms of savings in unit capital costs, for instance, the introduction of continuous casting, plus the concomitant higher yields, provided two-thirds of the capital savings associated with moving from 1960s to 1980s mini-mill technology. Continuous casting is also the most dramatic source of productivity improvements, although its predominance in this regard is less stark. In an integrated operation, by contrast, continuous casting affects only one segment of a much longer sequence of processes; its net effects are therefore less significant.

Furthermore, the billet casting employed in the production of wire rods replaces three other processes: ingot casting, bloom rolling, and billet rolling. The slab casting developed for flat-rolled products replaces only two: ingot casting and slab rolling. This difference implies that the overall savings provided by continuous casting will be greater for mini-mill products, an inference confirmed by the estimates of integrated wire rod costs (Table 7–7). The benefits of moving from 1960s technology to 1980s technology, in terms of both capital costs and labor requirements, are far greater for integrated wire rod production than for cold-rolled sheet: roughly 33 percent versus 17 percent in capital costs and 50 percent versus 33 percent in labor use. Thus, technological progress has been more robust in mini-mill products as well as in the mini-mill sector. The potential for improved performance in mini-mills, however, still exceeds the potential in integrated facilities, even for the same products. The effects of continuous casting are diluted in integrated plants even if the complex of flat-rolling facilities is excluded.

The benefits of simpler products are complemented by product specialization. This allows mini-mills to concentrate on being highly efficient producers of one or, at most, two related products; bureaucratic diseconomies of scale are minimized. By the same token, specialization allows mini-mills to gain economies of scale at the finishing end that are comparable to those achieved by integrated plants with much greater steelmaking capacity (spread over a wider range of products). Finally, specialization ensures that significant technological innovations have the greatest overall consequences for performance. As a result, the faster tempo of technological progress in mini-mill techniques has had spectacular cumulative effects. Technological improvements in one area (e.g., the electric furnace) have elicited corresponding advances in other processes (e.g., continuous casting and wire rod mills).

The technological and market–oriented dynamism of the mini-mill sector has ensured that high operating rates have been maintained, at the expense of integrated producers, and that high profits have been achieved. These features and results have characterized the spectacular growth of the mini-mill sector. Yet they can all be linked to the fundamental point that the entire mini-mill phenomenon represents the development of a capital-saving alternative to the integrated method of steelmaking. The American steel industry as a whole has faced the dilemma of attaining improved performance in the face of capital constraints, and the growth of the mini-mill sector has been its most successful response.

CONCLUSION

This chapter has shown that the traditional investment approach applied by integrated steel firms is no longer economically viable. The continued commitment to this approach—the massive-modernization strategy—is based less on a thorough analysis of the best means for improving performance than on a collective nostalgia for traditional ways of doing business, however inappropriate to present conditions. It is true that monumental facilities are marginally more efficient than facilities of adequate but smaller capacity, *ceteris paribus*. Yet the marginal benefits of massive scale count for naught if they come at the expense of operating rates or technological progressiveness. The effects of technology and high operating rates outweigh the effects of scale. A small, modern, and market-oriented plant, assuming it approximates the *minimum* optimal scale, will almost always outperform a larger but older plant.

Given the enormous cost of massive, Japanese-scale facilities, an attempt to compete along the dimension of scale is a sure recipe for low operating rates and lagging implementation of new technology, thus ensuring high costs and poor profitability.

Were the U.S. steel industry to evaluate its record of productivity growth, it would adopt an alternative strategy—one emphasizing the importance of capital-saving technical changes. Such a strategy would be almost the converse of the massive-modernization strategy. Whereas the massive-modernization strategy assumes that improved performance is impossible without major capital expenditures, a more appropriate strategy would seek continuous marginal improvements in operating techniques. Whereas the massive-modernization strategy only weakly discriminates among various capital projects and in general emphasizes scale increases, a more appropriate strategy would target technological breakthroughs and restrict scale to the minimum necessary to gain adequate economies. This would also make it easier to maintain high operating rates, which are crucial to financial performance. Whereas the massive-modernization strategy has an insatiable appetite for investment funds, a more appropriate strategy would seek investments that reduce the burden of capital costs. This is a theme to which we shall return in Chapter 10.

Notes

1. Winston Churchill, *The Second World War; the Gathering Storm* (Boston: Houghton Mifflin, 1948), p. 475.
2. American Iron and Steel Institute (AISI), *Steel at the Crossroads* (Washington, D.C.: AISI, 1980) and *Annual Statistical Report, 1982* (Washington, D.C.: AISI, 1983).
3. Throughout this chapter, estimates are used that are based partially on published sources (including statistical publications and industry studies) and partially on data provided by steel firms and equipment suppliers. The principal published sources are listed at the end of Appendix C. Besides these, data provided by the following companies were particularly helpful: Union Carbide, Whiting, Concast, Ferrco, Demag, Rouge Steel, Stelco, Dofasco, and Raritan.
4. The capital costs of an integrated plant producing cold-rolled sheet are representative. Hot-rolled sheet would cost somewhat less (about \$900/ton) since it does not require some finishing operations. For the opposite reason, coated products such as galvanized sheet and tinplate would entail higher capital costs (about \$1550/

ton). Capital costs for a plate mill would be about $750 per ton, while most pipe and tube would cost considerably more than cold-rolled sheet.

5. We assume that the integrated plant producing wire rods is devoted entirely to barlike products (bars, rods, and light structurals)—even though such a plant is unlikely to ever exist. Flat-rolled capacity would raise unit capital costs.
6. A similar argument has been made by Robert Crandall, *The U.S. Steel Industry in Recurrent Crisis* (Washington, D.C.: The Brookings Institution, 1981), pp. 72–92.
7. In constructing this table, we assume that the Japanese industry enjoys its average advantage in operating techniques in the categories of packaging, shipping, and overhead, even though more labor is required for these operations as a result of economic choices; hence there are entries for these process steps under both economic choices and operating techniques.
8. For example, more process steps are eliminated for this product (bloom rolling and billet rolling as opposed to slab rolling alone). Moreover, the transition to continuous casting involved the replacement of the least efficient primary-rolling facilities. Ingot tailoring and other practices also contribute to this result.
9. For both mini-mills and integrated plants, continous operation has required multiple furnaces and casters, increasing the minimum optimal scale. However, mini-mills are now devising ways to minimize the problems associated with single furnace operations (e.g., Raritan's plant). This can reduce the minimum optimal scale to well below 750,000 tons.
10. At small scale in products where economies of scale are significant (e.g., at less than two million tons for cold-rolled sheet), the capital-labor ratio initially declines before rising.
11. It is worth emphasizing in this context that managerial attitudes, as well as technological factors, are crucial to the lower capital costs of mini-mills. When mini-mill facilities are built by integrated firms, their capital costs are often well above the mini-mill norm, since they lack the "build cheap, build quick, and build tight" philosophy of the mini-mills. This is particularly characteristic of the electric furnace plants, often including direct reduction units, built in developing countries.

III POLICY AND PROSPECTS

8 THE DILEMMA OF INDUSTRIAL POLICY: LESSONS FROM ABROAD

"We called him Tortoise because he taught us," said the Mock Turtle angrily.[1]

—Lewis Carroll
Alice in Wonderland

We now turn to the role of government policy—a theme that, although crucial, has been kept in the background until now. It is important to resist the temptation to blame the American steel industry's postwar performance on an unsympathetic government. Yet it would also be incorrect to ignore the fact that the government largely establishes the environment in which market forces operate. While this is widely accepted in regard to macroeconomics, the impact of political actions on individual sectors of the economy may be even more definitive than is the case at the macro level. In this chapter and the next, we will focus on the interaction between the steel industry and the government in the United States, Japan, and Canada.

Besides its importance for steel, an evaluation of the governmental role in, or response to, the declining competitiveness of the American steel industry is an essential element in the broader debate about the future course of governmental involvement in the U.S. economy. Stagflation and lagging international competitiveness have encouraged some politicians or interest groups (e.g., the AFL-CIO) to call for a more active government role in allocating capital or restructuring economic activity.[2] Such an approach, generally referred to as industrial policy, seems to characterize the industrial organization of several of our more successful competitors—particularly Japan. Unfortunately, in the United States there is little agreement concerning the key features of the industrial policy approach, so that it is more a political slogan than a set of realizable programs.

Steel policy is an outstanding test case for determining the potential and limits of an industrial policy approach. Steel has played a central role in the strategic industrial planning of every country that has applied this approach, and it has also been the target of many of the halting attempts to implement industrial policies in the United States. In this chapter we will discuss how government has nurtured the steel industry in two countries where this has been highly successful: Japan, the generally preferred model for advocates of industrial policy, and Canada, a less evocative but perhaps more fruitful model due to its cultural similarity to the United States. Chapter 9 will then turn to the complicated relationship between governmental policy and the American steel industry.

JAPAN: FORCED MARCH TO WORLD LEADERSHIP

The success of the Japanese steel industry represents one of the best examples of how a supportive policy environment can nurture an industry. Just as importantly, it confirms the view that the right mix of government policy and industry strategy can drastically alter the structure and dictates of "market forces." The present superiority of the Japanese steel industry is an undeniable fact; therefore, economic analyses of the reasons for this result generally cite a roster of economic factors that have "caused" Japan's present competitive superiority and that retrospectively appear quite obvious. Lower raw materials costs, lower labor costs, higher productivity, larger plants, and so on—these are some of the advantages that have propelled the Japanese industry into first place among the world's steel producers.

In the late 1940s and early 1950s, however, any suggestion of the future success of the Japanese steel industry would have seemed bizarre indeed. Located in a country almost completely devoid of raw materials and energy sources, the small Japanese steel industry had older, smaller plants. Its labor productivity was abysmal, and it was forced to rely on capital markets that were small and undeveloped. In fact, the only clear "advantage" enjoyed by the Japanese steel industry was cheap labor, a traditional characteristic of underdeveloped countries. One would have been hard pressed at that time to define the structure of comparative advantage that has led to the Japanese steel industry's competitive prowess and that in hindsight seems so clear-cut.

The key to the rapid emergence of the Japanese steel industry in the postwar period lies in the successful interaction of three factors: positive economic conditions (e.g., the potential of rapid market growth), appropriate managerial strategies, and a supportive policy environment. In the early postwar period, when both the overall economy and the steel industry were weak, the government played the key role. This was clearly the case during the American occupation, which lasted until 1952. Government policies were also important during the 1950s, the decade of the modernization programs that progressively overcame the competitive disadvantages faced by Japanese steel producers. Success gradually reduced the need for activist government support, however, so that by the 1960s the industry became the dominant pole of the government-industry consensus.

Because of the devastation caused by World War II and the impact of the American occupation, the pattern of economic growth in the first years after the war was largely determined by the state, which controlled most of the available capital resources as well as the flow of trade. Even before reconstruction was complete, the government designated infrastructural sectors as the strategic or core industries: coal, steel, shipbuilding, electric power, and fertilizers. Among these, steel and coal were given pride of place; this ensured that they would receive a greater share of resources than "market forces" would otherwise have dictated. Strategic industries were favored with incentives on every front, both directly (e.g., through subsidized investment and tax benefits) and indirectly (e.g., through state-sponsored rationalization and consolidation). They were protected from foreign competition, and they attracted the country's elite university graduates.

The strategic concentration on steel and coal was not based on a static identification of Japan's comparative advantages. Given the limited resources available for reconstruction, the identification of steel as a strategic industry provoked some debate within political circles. Several influential bankers and politicians sought alternative goals for industrial policy, believing that Japanese steel producers had no real prospects for competing in the world market, particularly against the American industry.[3]

This resistance was based on the severe obstacles that confronted Japanese steel producers in the first years after the war. The principal barrier, one shared by all Japanese industries, was financial; in the desolation of defeat, private capital was wholly inadequate to finance repair, expansion, or modernization. This problem was compounded by the dearth of domestic raw materials—in sharp contrast to the seemingly easy access to raw materials enjoyed by U.S. and European producers. Furthermore, importation of raw materials required foreign exchange, something that no Japanese industry possessed in adequate amounts. Finally, Japanese costs of production were extremely high by world standards; low labor costs were offset by abysmal labor productivity due to the inferior technology of Japanese plants. Japanese steel producers simply could not compete in unprotected markets without some form of government support.

Aggressive governmental intervention alleviated or eliminated these disadvantages, particularly capital constraints, in the immediate postwar period. Chiefly through the government-owned Reconstruction Finance Bank, the state provided fully one-half of the Japanese steel industry's funding from 1946 to 1948.[4] Second, governmental subsidies were provided to reduce the price of raw material imports, especially iron ore. Low-priced steel was viewed as crucial to industrial expansion, but this was provided via a massive program of price subsidies rather than price controls. This guaranteed both high revenues for steel producers and low prices for steel consumers. In July 1948, raw materials subsidies (e.g., for iron ore) and price subsidies in the steel market covered an incredible 76 percent of steel firms' costs.[5] Finally, import barriers ensured that the recovery and expansion of Japanese steel firms would not be undermined by low-cost foreign competition—especially from the United States.

In the critical years from the end of World War II to 1950, the government fostered a delicate but real dynamism in the Japanese steel industry. Once prosperity gradually returned, augmented by the Korean War, the industry began to develop its own strategy for overcoming its inherent disadvantages. Assured of adequate access to financing and subsidized imports of raw materials, the industry's immediate priority was to overcome the high costs that limited its participation in world markets. From the outset, Japanese steel producers defined their strategic goal in terms of international standards, which in the 1950s clearly meant emulating the American industry. Cost competitiveness required modernization of the Japanese industry's antiquated facilities, and that was the focus of the First Modernization Program, in effect from 1951 to 1955.

This program, an expression of the industry's strategy, was administered through the government's Ministry of International Trade and Industry (MITI), then emerging as the central organ of Japanese industrial policy. As in the 1940s, the government's principal function in the First Modernization Program was financial support. Various government sources provided roughly 40 percent of the industry's funding during this period.[6] Tax incentives were also used to foster modernizing investments. Tariffs were lowered on imports of equipment, especially from the United States, and conscious efforts were made to learn from and adopt American techniques. The government also sought to rationalize the transportation networks that served the industry, especially to facilitate raw materials imports. Finally, the First Modernization Program explicitly defined the government's role in investigating and guiding the industry's efforts. As far as the industry was concerned, it focused its investments on finishing operations rather than on expansion. Efficiency was raised to the point where Japanese steel prices had more or less fallen to international levels by the end of the program.

In the ten years between the end of World War II and the end of the First Modernization Program, the interaction between government policy and industry strategy achieved stunning success. Milestones were passed in rapid order: record peacetime output was achieved by 1951, record output by 1953, and price competitiveness in the home market by 1955. Each of the obstacles that had confronted the industry in 1945 had been overcome: government subsidies had alleviated disadvantages on the capital and raw materials fronts, while the industry's investment program had generated enormous efficiency improvements–although competitiveness was still based on low wage rates. By the end of the First Modernization Program, the strategic problems faced by the Japanese steel industry had shifted. A more dynamic or long-term resolution of the industry's competitive disadvantages was now required–one that would be based more on economic performance than on government assistance.

These conditions determined the content of the Second Modernization Program, which lasted from 1956 to 1960. Japanese steel producers were now poised to compete successfully in the international steel market, and domestic demand was booming. By the end of 1955, operating rates were so high that export controls were imposed to ensure that the Japanese economy had an adequate supply of favorably priced steel. Against this background, the Second Modernization Program emphasized the construction of new plants; steel capacity almost tripled during this period. Investment in the Second Modernization Program was more than four times as great as during its predecessor. Government sources now accounted for only 15 percent of total investment, although international lending bodies like the World Bank provided another 12 percent.[7] Substantial tax benefits were maintained, however, as were controls on imports. During the Second Modernization Program, Japanese steel producers continued to rely on foreign technology, especially from the United States–although by the end of the program the pupil was beginning to outstrip the master. As late as 1961, over 60 percent of the Japanese industry's sales were dependent on imported technology; by 1967, this had fallen to 8 percent.[8]

Throughout the period of the modernization programs, government policies had been determined within the context of a fruitful consensus between the industry and the government. As the government's role receded, the dynamic focus shifted to the industry's strategy. Performance improvements were sought not by the replacement of outmoded facilities but rather by the construction of greenfield plants at tidewater sites. New Japanese facilities soon defined the world standard of efficient scale. Investments were concentrated on the hot end, since greater efficiency in the furnaces reduced the impact of higher priced raw materials. It was a highly felicitous circumstance that this first burst of expansion coincided with the commercialization of the basic oxygen furnace (BOF). Besides striving for operating improvements, Japanese producers sought long-term contracts with raw materials suppliers—an effort that was particularly beneficial in regard to iron ore. Very early on, the Japanese turned toward new sources of iron ore in Australia and South America. Huge bulk carriers were built to transport these materials, reducing unit shipping costs; and tidewater sites ensured that handling charges were minimal. These measures transformed the initial disadvantage of poor domestic resources into a competitive advantage. By the 1960s, European and American reliance on inferior domestic ores, frequently transported by less cost-effective means, had become a handicap.

More attuned to the implications of the increasing internationalization of markets in minerals and steel, the Japanese were more adept at positioning themselves to exploit this trend. Massive greenfield plants built at tidewater brought benefits on many fronts: increasing labor productivity, raising efficiency in raw materials use, lowering the delivered price of raw materials, and reducing the cost of export.

By 1960, the Japanese steel industry had established itself as one of the foremost in the world, even if the immensity of its achievement was still underestimated. Even more spectacular successes would follow in later years, but by this time the industry's expansion and improvement were basically self-sustaining. Overt governmental support, especially in terms of funding, therefore receded, although tax benefits and trade barriers persisted. Increased international competitiveness—a sine qua non for a country as resource poor as Japan—remained the overriding goal for both the industry and the state. By this time, however, the industry's competitive strategy (which had been developed with substantial government input) was the principal guarantor of success. No longer needed as an instigator, the state became more of a facilitator and arbitrator.

Nevertheless, the structural features of the Japanese steel industry continue to reflect the period of governmental activism. For instance, internally generated funds still play a small role in investment. Debt-equity ratios in the Japanese steel industry are extremely high by American standards; and while the government is no longer the principal source of funds, it acts as the implicit guarantor of loans to the industry through the commitments of the Bank of Japan and the network of relationships among the state, the banks, the trading companies, and the manufacturing sector. The highly leveraged status of Japanese steel firms, a legacy of the early postwar period, is a significant determinant of their behavior.

The Japanese government now plays primarily an advisory role, striving to ensure that the plans of individual firms reflect the industry-government consensus and constitute a coherent industry strategy, especially during periods of weak demand. The government is expected to minimize the disruptions caused by business cycles—not just in terms of macroeconomic policy, as in the United States, but through industrial policy as well. Under normal market conditions, intense competition among domestic producers is encouraged as a means of raising efficiency. During recessions, however, price-fixing arrangements are tolerated or even encouraged. Such intervention may be vital in Japan, since the highly leveraged balance sheets of Japanese steel firms leave them exceedingly vulnerable to reduced revenues and thus exceedingly prone to self-defeating orgies of discounting when demand lags.

Even during better times, however, competition can be restrained by the state when it threatens to weaken overall competitiveness, as this is defined by the industry-government consensus. Since steel profitability is greatly affected by operating rates, the tempo of investment and expansion by each individual firm is a matter of industrywide concern. During the modernization programs, government funding ensured that the projects undertaken by individual firms reflected a collective assessment of what was needed. Since that time, the consensus has generally been enforced by subtler means—a pattern facilitated by the close links among government officials (especially in MITI), financial institutions, trading companies, and steel firms. Should a maverick firm violate the consensus, MITI has several means of discipline at its disposal. In 1965, for example, the ministry used its control of import licenses to limit the supply of coking coal to Sumitomo, the most aggressive and dynamic of the Big Six Japanese steel firms. Despite an industry-government consensus (largely built around the views of the more traditional steel firms) that expansion should be deferred, Sumitomo had gone ahead with plans to expand and upgrade its rolling capacity. By 1967, a strengthened market had rendered the issue moot; the incident nevertheless indicates the extent to which the state is willing to intervene and even punish firms that fail to accept the prevailing wisdom.

Intervention was rarely needed in the 1960s and early 1970s, but the governmental role has increased since the onset of the world steel crisis in 1975. The emergence of excess capacity (Japanese operating rates have averaged roughly 65% since 1975) has brought the growth years to an abrupt halt. A rather strict price cartel is now evident, along with a moratorium on expansion and an orderly program of mothballing relatively inefficient facilities. Investment is still high by world standards, but it is oriented toward raising efficiency, improving quality, and solidifying Japanese producers' competitiveness in the high-value products that are less vulnerable to competition from developing countries. There is a growing consensus that the Japanese steel industry is now entering a stage of decline, despite its undoubted preeminence in the world market. The government seems to be assuming a larger role in attempting to manage this process of decline, just as it was very active in the first stages of growth.[9]

This cursory history of postwar Japanese steel policy has been highly selective; it has obviously ignored the contradictory and ad hoc elements that would be given a prominent place in a more thorough treatment. At the very least, it must be emphasized that the actual process of formulating and implementing steel policy in Japan is less smooth than this description may have implied. Nevertheless, it suggests the extent to which governmental policies have nurtured the expansion of the Japanese steel industry. More significantly, we have sought to indicate the state's integral role in this process; its importance extends far beyond the passive establishment of a favorable policy environment to which firms merely adjust.

Nonetheless, it is actually industry rather than the government that dominates the Japanese system of industrial policy—if such a distinction can even be made. In terms of steel policy, the process of developing a consensus to guide the performance (e.g., investment planning) of individual firms begins with the Japan Iron and Steel Federation (JISF), the industry's trade association. This consensus is then refined through the explicit involvement of MITI officials, after which it is formalized in official planning. MITI can seek to enforce its own views, but instances of diktat are exceedingly rare. Even the controlled adjustment to decline—presumably an essential feature of a successful long-term industrial policy—can be undertaken only when two-thirds of an industry requests it.[10]

Finally, this assessment of Japanese steel policy must end with the warning that it is easy to exaggerate the permanence of Japanese success. Even a relatively bumbling set of policies can be blessed with the appearance of success if their effects are offset by positive economic conditions and an effective industry strategy. It is relatively easy to manage growth, but the skills required to overcome decline or stagnation are quite different. That is the task that now faces the Japanese steel industry and Japanese steel policy, so that breakdowns in the strategy-policy consensus have become more likely. Indeed, MITI has had misgivings about the sale of Japanese steel technology to developing countries such as Korea and Brazil, since this creates new competitors in the long run. Given their highly leveraged positions, however, Japanese steel firms have sought revenues from all sources to offset the lagging demand for their finished product. This is one example of how altered competitive prospects can unravel a once impressive record of consensus.

CANADA: MODERATION IN ALL THINGS

As in Japan, the postwar success of the Canadian steel industry was grounded in a progressive consensus between industry and the state—a consensus that grew out of a shared perception of the strengths and weaknesses of the Canadian economy. Policy goals were established in terms of international competition and were shaped according to the political and historical conditions that define the role of the Canadian government.

Public Policy and the Canadian Economy

In spite of a radically different history, the Canadian political system, like that of the Japanese, fosters a consensus approach to policymaking. Canada is a parliamentary democracy; since there is only one elective body (the House of Commons), and since the same bloc or party must control both the legislative and executive functions, there is potentially less uncertainty in Canada than in the United States as to the role and ends of government policy. Canada's constitution has not been formalized or amended via strict legal formulas, as has been the case in the United States. Canadian notions of common law, based on the presumption of broad social agreement, replace constitutional strictures.

The Canadian political system thus assumes that consensus plays a larger role in forging policy than is the case in the United States. Rather than being detailed by statute, the behavioral guidelines for Canadian officials are broad and flexible. In principal, this approach is more consistent and effective than the system of trial by conflict (in the courts, in political campaigns, in the media, etc.) prevalent in the United States. Governmental sensitivity and effectiveness are also facilitated by the traditions of the Canadian civil service, where even high-level positions are career appointments rather than short-term political sinecures. This idealized image, however, should not be exaggerated. The effectiveness of the Canadian mode of policymaking rests on a fragile basis; it is highly dependent on the programs and leadership of the politicians in power. Should the political element seek to dominate and thus destroy the consensus, the legal apparatus offers less protection than does the American system.

The policy goals that drive this political system, at least in terms of economics, are defined by the physical and historical features of the Canadian economy. Canadian markets are widely dispersed, and the centrifugal tendencies inherent in this geography are intensified by the fact that Canada was patched together from separate British colonies—without the unifying experience of a political revolution. Hence, the development and maintenance of a common economy has always been an imperative of Canadian nationhood. Internally, this goal was sought by governmental development of a national transportation system (especially in railroads) and by the development of a federal financial network. The external complement to this was the erection of tariff barriers to create internal trade and to prevent each Canadian region from trading separately with foreign countries, especially the United States and Great Britain.

The key element in the effort to develop a truly Canadian economy was the perception that Canada is resource-rich; labor and capital are scarce. In an effort to avoid becoming solely a source of raw materials for more developed economies, Canadian policymakers have traditionally sought to use high tariffs on manufactured goods to create a domestic manufacturing sector that could process indigenous raw materials. Unfortunately, this goal has been constrained by the limited size of the Canadian markets; costs have therefore been high by U.S. standards. Furthermore, the scale of the U.S. market and the structure of U.S. tariff barriers—higher on manufactured goods than on raw materials—have acted as counterweights to Canadian efforts to encourage manufacturing and to upgrade raw materials, so that a large share of Canadian resources have traditionally been processed in the United States.

Due to these conflicting pressures, traditional Canadian tariff barriers had somewhat perverse results. The tariffs enticed U.S. companies to locate branches in Canada, distributing products designed in the United States but assembled in Canada—a pattern encouraged by the existence of a common language and by proximity to common communication systems. U.S. ownership proliferated in Canadian manufacturing and resource extraction. Conflicting policies and market forces in the United States and Canada thus exacerbated the competitive weaknesses of Canadian manufacturers, since protectionist policies led to a proliferation of domestic and foreign-owned firms vying for the small Canadian market.

The disappointing consequences of the rather crude policy of high tariffs led to the development of more sophisticated incentives after World War II—a trend that was encouraged by the General Agreements on Tariffs and Trade (GATT), which reduced international trade barriers. Greater involvement in the world economy elicited a convergence of interests between industry and the state, a convergence that was particularly fruitful in the steel industry. Still based on the perception of Canada as resource-rich, Canadian policies were designed to promote the extraction and upgrading (at least to the smelted level, e.g., pig iron) of raw materials and the growth of primary manufacturing. The incentives employed were the familiar tools of tax policy: both a generous allowance for resource extraction and processing and an accelerated schedule of capital recovery (e.g., manufacturing equipment could be written off in approximately two years) were instituted in the 1950s. These measures deferred tax payments by Canadian firms, increased cash flow, and thus provided some compensation for the high Canadian tax rates which had traditionally financed extensive state involvement in the economy, including ownership in such sectors as power generation, railroads, electronic media, and so on.

The Canadian Steel Industry in the 1950s

Despite high tariffs, Canada has traditionally imported large amounts of steel from the United States, since transport costs are relatively low and the small Canadian market limited the scale and efficiency of Canadian plants. Nonetheless, steel was not protected to the same extent as was the manufacture of finished products, so there was less incentive for American steel firms to build facilities in Canada. As a result, Canadian steel plants are domestically owned, while U.S. ownership has been much more prevalent in raw materials extraction (e.g., iron ore) and in the manufacture of finished products (e.g., automobiles). Canadian steel production is heavily concentrated in southern Ontario, one of the largest regional markets, drawing coal from West Virginia and iron ore from both U.S. and Canadian mines.

In the early 1950s, the Canadian steel industry was an inefficient, minor player in the world steel market.[11] This is suggested by Table 8-1, which compares Canadian and U.S. performance in 1958 and 1981. By 1958, the benefits of the emerging steel consensus between the industry and the government were already apparent. Nevertheless, even at the end of this decade, Canadian imports were more than twice the level of exports. Unit costs were high for each of the major input categories. Despite the perception that abundant raw materials provided the Canadian steel industry with its comparative advantage, no such advantage is evident in Table 8–1. Lower Canadian employment costs were needed to offset a still significant disadvantage in labor productivity. Finally, capital costs have always been higher in Canada than in the United States; although this is not shown explicitly in the table, it is reflected in the comparison of total costs. During the 1950s, Canadian competitiveness was based largely on lower labor costs—as was also the case with the Japanese steel industry.

Table 8–1 also describes the same comparative measures for 1981. The contrast is striking; in many areas (particularly trade flows) the status of the two industries has been reversed. The weak Canadian industry of the 1950s had by 1980 transformed itself into one of the most efficient in the world, while the United States had long since fallen from its position as the world leader.

Table 8–1. U.S. And Canadian Performance: 1958 versus 1981.

	1958		*1981*	
	Canada	*U.S.*	*Canada*	*U.S.*
Imports/Apparent consumption[a]	24.7%	3.3%	15.9%	17.6%
Exports/Total shipments[a]	9.8%	4.7%	24.4%	3.5%
Average integrated plant capacity[b]	1.3	2.5	4.0	3.8
Productivity (MHPT - CRS)	12.5	11.5	6.3	7.0
Hourly employment costs (US$)	2.75	3.75	12.75	20.15
Iron ore costs (cif - US$/ton)	10.50	10.64	38.50	40.00
Coal Costs (cif - US$/ton)	11.00	10.50	61.00	57.50
Coke rates	0.75	0.80	0.43	0.55
Estimated Production cost (US$/ton of CRS)	120.00	122.00	355.00	445.00
Price (US$/ton of CRS)	130.00	130.00	385.00	460.00

a. Five-year average.

b. In millions of net tons.

Sources: *Statistics Canada* (various years); American Iron and Steel Institute (AISI), *Annual Statistical Report* (New York and Washington, D.C.: AISI, various years); World Steel Dynamics, *Core Report J* and *Core Report Q* (New York: Paine Webber Mitchell Hutchins, 1979 and 1982, respectively); Prices and Incomes Commission, Canada (various studies, especially *Steel and Inflation,* 1970); and various studies done by D.F. Barnett for U.S. and Canadian governments.

The Dynamics of Change: 1955–1975

While no explicit steel policy was promulgated by the Canadian government, the Canadian experience exhibited the same key element that was evident in Japan's more formal industrial planning, namely, close cooperation between industry and government. The roots of this relationship can be discerned in the 1940s, when a common view began to emerge around the issue of raising Canadian efficiency to U.S. levels. This goal became even more pressing when the GATT required the dismantling of the tariff barriers that had formerly been the centerpiece of industrial policy in Canada. Expanded competition in world markets made it essential that Canadian manufacturers be assured of an adequate supply of low-priced steel—a consideration that provided the government with an incentive to encourage greater efficiency within the industry. Moreover, the perception that Canada had a comparative advantage in raw materials encouraged the view that it could compete in industries that, like steel, were closely linked to raw materials extraction. Drawing on the existing structure of the Canadian steel industry, by the mid-1950s a progressive consensus had been forged concerning the best means of overcoming the obstacles that had traditionally blocked the emergence of a world-class steel industry in Canada.

The crucial weakness of the Canadian steel industry had traditionally been the small size of its home market and the concomitantly small scale of its facilities. Because of this, Canadian producers had always sought to emphasize specific products, and this pattern was encouraged by the policy-strategy consensus in effect from the mid-1950s to the mid-1970s. The benefits of specialization were recognized by the government, and antitrust principles (termed anticombine in Canada), which prohibit even a tinge of such collusive allocation of markets in the United States, were applied in such a way as to allow the allocation of markets among different firms. Faced with the disadvantage of a small domestic market, both the industry and the government favored increased specialization in order to gain economies of scale and to increase Canada's international competitiveness. This strategy-policy consensus, which minimized interfirm jockeying for market share, lowered the overall risks of steel investment.

The second major disadvantage faced by the Canadian steel industry was the relatively high cost of capital—a weakness that had retarded growth and technical progress. The interaction of policy and strategy in overcoming this obstacle was more complex. One element was a drastic change in overall tax policy, which boosted the cash flow of industries like steel by reducing the depreciation period to roughly two years (compared to over fifteen in the United States). This served to encourage investment. Less visible, although just as important, was the managerial concentration on maintaining high operating rates, which had the effect of decreasing the overhead costs associated with a given level of output. The tacit consensus was that Canadian steelmakers should build sufficient capacity only for the markets in which they had or could obtain a comparative advantage, an advantage that was partially due to tariffs and transport costs. Much of the Canadian market was thus left to imports, which provided a cushion to Canadian producers. When market conditions worsened, Canadian steel firms could then "dump" in those domestic markets normally supplied by foreign producers—a practice facilitated by the government's purposeful application of the trade laws.

The maintenance of high operating rates became the keystone of the managerial strategy of Canadian steel firms. The application of this strategy in effect transformed the substantial import penetration of the 1950s from a liability into a strength. The significance of high operating rates was discussed in Chapter 7; in the Canadian case, their importance is suggested by Table 8–2. The way in which trade flows facilitated the maintenance of high operating rates is indicated by the data on market fluctuations. Canadian shipments grew more rapidly and were less variable than domestic consumption between 1958 and 1980. U.S. shipments were more volatile—a difference that became more pronounced over time. The data provided on market dampening show that during downturns the Canadian steel industry tended to replace imports and boost exports, while imports rose, and exports fell, when demand was high. This dampened the effects of market cycles on operating rates by an average of almost 25 percent. In the United States, on the other hand, the effects of market downturns on domestic shipments were actually exacerbated; imports replaced domestic shipments rather than the reverse. This is shown by the large negative values for relative dampening and the comparatively low U.S. operating rates.

The strategy-policy consensus in operation from the mid-1950s to the mid-1970s was founded on the commitment to specialization and high operating rates. Policymakers accepted the exigency of reduced domestic competition via product specialization and the necessity of facilitating more rapid capital formation. The government had a thorough, ongoing grasp of how the system worked and of the need to nurture what was in effect an infant industry; as a result, rules (e.g., anticombine and environmental) were flexibly enforced. Tight domestic supplies were sometimes an unwanted byproduct of the effort to raise operating rates, and moderate levels of imports were viewed as an antidote to this potential brittleness. Not having overbuilt, the Canadian industry was less exposed to the vagaries of the world market and was therefore better able to benefit from supportive government polices. As their part of the tacit

Table 8–2 Capacity Utilization, Market Volatility, and Performance: U.S. versus Canada, 1958–81 (annual percentages).

	1958–64		*1964–72*		*1972–81*		*1958–81*	
	Canada	*U.S.*	*Canada*	*U.S.*	*Canada*	*U.S.*	*Canada*	*U.S.*
Capacity utilization (effective)	84.4	68.0	89.3	84.5	89.2	83.8	87.8	79.5
Variations in apparent consumption[a]	±7.5	±4.8	±4.8	±3.5	±7.3	±8.7	±6.5	±5.9
Variations in total shipments[a]	±6.0	±5.8	±4.0	±4.9	±4.1	±8.3	±4.6	±6.5
Market dampening: absolute[b]	+1.5	−1.0	+0.8	−1.4	+3.2	+0.4	+1.9	−0.6
relative[c]	+20.0	−20.8	+16.6	−40.0	+43.8	+0.5	+29.2	−10.2
Return on sales	9.9	5.5	8.9	4.6	7.8	3.6	9.0	4.4

a. Variation in each year's consumption or shipments from respective five-year moving average.
b. Apparent consumption variations minus total shipments variations.
c. Apparent consumption variations minus total shipments variations, divided by apparent consumption variations.
Sources: As in Table 8–1.

bargain, the companies ensured that price increases were moderate, that the government was kept informed of developments within the industry, and that regulatory standards were met or exceeded, so that no unnecessary embarrassments occurred.

The emergence of this consensus was eased by favorable developments within the Canadian economy. During the 1950s and 1960s, the tempo of raw materials extraction increased. Infrastructure was built up rapidly, urbanization increased, and there was a strong impetus for the expansion of manufacturing facilities. Thus, the demand for steel increased at an annual rate of 5 to 6 percent from the mid-1950s to the mid-1970s (see Table 8–3).

Planted in this environment, the steel consensus between government and industry soon bore fruit. Using the favorable tax policies, Canadian steelmakers quickly expanded—but did not overexpand—to meet market growth, so that modern new facilities came onstream at a fairly rapid pace. Newer, larger, and more advanced facilites created an efficient and competitive industry; this in turn expanded the markets in which Canadian producers were competitive. Imports were gradually replaced, so that growth in domestic shipments outstripped market growth. Capacity utilization remained high, and eastern and western Canada, traditionally supplied by imports, became in effect a dumping ground for Ontario producers during market downturns.

Managerial strategies of specialization and modernization developed rapidly under these conditions. During the 1960s, some product lines traditionally imported into Canada were taken over by domestic integrated firms, but several products were spun off from integrated producers to mini-mills. Unlike their U.S. counterparts, Canada's integrated mills did not seek to compete with mini-mill operations in barlike products; instead, they either surrendered these markets or built their own mini-mills. Both types of response facilitated the overall strategy of specialization. Besides increasing the efficiency and profitability of Canadian integrated producers, specialization was also a great boon to Canadian mini-mills. Several Canadian mini-mills (e.g., Ivaco and Ferrco) became world leaders in mini technology and were able to use this advantage to expand not only in Canada but abroad, especially in the United States.[12]

By the 1970s, the great majority of Canada's merchant bar, wire rod, and light structural production was carried out by mini-mills. There were three major integrated producers, all with modern, efficient facilities. One of these firms (Algoma) concentrated on structurals, rails, and plates; another (Dofasco) almost exclusively emphasized narrow flat-rolled and coated products. Only Stelco, the third integrated firm and the industry leader, produced a range of products at all comparable to

Table 8–3. Comparative Performance: U.S. versus Canada, 1958–81 (annual percentages except where noted).

	1958–64		*1964–72*		*1972–81*		*1958–81*	
	Canada	*U.S.*	*Canada*	*U.S.*	*Canada*	*U.S.*	*Canada*	*U.S.*
Growth in apparent consumption[a]	5.7	3.4	6.2	2.8	0.9	1.2	3.9	1.4
Growth in total shipments[a]	8.4	2.2	5.6	2.1	2.7	1.7	5.1	0.6
Capital expend. (US$/ ton shipped)	26.00	16.88	24.00	21.55	41.50	33.50	31.15	24.67
Productivity improvement[b]	5.5	3.1	3.1	3.2	1.9	0.8	3.3	2.3
Return on equity	12.1	7.5	10.4	7.1	12.8	8.7	11.9	7.7

a. Calculated using five-year averages around year listed.
b. Raw steel production divided by total manhours worked (all products). Exaggerates improvements for any one product because of changes in product mix (bias greater for United States than for Canada).
Sources: As in Table 8–1.

the product mix characteristic of U.S. integrated producers. Furthermore, until the late 1970s, when Stelco built the newest integrated plant in North America, each integrated firm had only one major steelmaking plant. Average plant size had reached the U.S. level, so that scale was no longer the relative disadvantage it had once been. Canadian firms set world standards for some technologies, especially in regard to improvements in operating techniques (hot-metal desulphurization, direct rolling, coil boxes, etc.).

By the late 1960s, Canadian producers had reached competitive parity with the U.S. industry in many product lines—a result caused by several factors, for example, the lower value of the Canadian dollar and the rapid reinvestment of funds in ever more modern facilities. Many of these factors were conditioned by the policy-strategy consensus discussed above. By the 1970s, Canada was cautiously exporting ever greater amounts of steel to the United States, as the natural market in which Canadian output was competitive expanded to include the entire Great Lakes region.

By the late 1970s, the policy-strategy consensus that had been in force from the mid-1950s to the mid-1970s had produced results that, although less spectacular, in many ways matched the achievements of the Japanese steel industry. During the last decade, the Canadian steel industry has been the most profitable in the world (Table 8–3). It has emerged as a sizable net exporter, reversing its traditional status. By 1975, its costs and prices were far below U.S. levels, its plants were modern and efficient, and it was planning for a promising future.

Cracks in the Consensus

The fragility of the Canadian "consensus" mode of policymaking has been evident in recent years. From 1955 to 1975, the strategy-policy consensus had been built on a delicate foundation of communication and sensitivity, entailing a high degree of flexibility on the part of government officials. By the early 1970s, government intervention became more erratic, although in some cases this benefited the steel industry. Wage and price controls were established on several occasions; at times these coincided with contractual negotiations in the steel industry, so that labor costs were restrained. This contributed to increased competitiveness; by the late 1970s, employment costs in the Canadian steel industry were about 30 percent above the manufacturing average, as compared to a differential of about 70 percent in the United States. Other governmental actions, however, had a negative impact on steel firms. Several well-publicized investigations of steel price increases were instituted, for instance. The industry was "exonerated" in each case, and these investigations never extended into threats to alter the structure of the industry for the sake ofgreater product competition.[13] Nevertheless, day-to-day operations were disrupted. It was clear by the early 1970s that the mode of Canadian policymaking was changing.

Events in the steel industry reflected overall trends in the Canadian economy. The broader national consensus began to disintegrate under the pressure of declining growth, regional rivalries, and a more active stance on the part of the government. The gradual disappearance of consensus politics was accompanied by the proposal of a more explicit industrial policy. The identification of Canada's comparative advantage with an abundance of raw materials became more pronounced, and policies were designed to exploit this alleged advantage. This represented an extension of previous policies. In earlier years, however, the advantage of abundant raw materials had been merely assumed. In the 1970s, policy became more directive: the exploitation of this resource advantage was now enforced. Unfortunately for steel producers, the alleged comparative advantage in the raw materials of steelmaking is an illusion, except in the minor area of electricity costs (for mini-mills). Canadian iron ore, for all its abundance, is not an advantage for the domestic steel industry when compared with richer, lower cost deposits in Brazil, Australia, and elsewhere (see Appendix B).

Due to the misconceived notion of a resource-based Canadian advantage in steel production, pressure built for emulating the Japanese strategy by building additional capacity targeted toward the export market. Furthermore, the periodic steel shortages that were provoked by the policy/strategy of maintaining high operating rates (not overbuilding) also increased pressure for expanded capacity. Finally, increased economic contention among provinces led to competition for new steel capacity and to provincial government subsidization of outmoded facilities. The success of the Canadian steel industry had greatly depended on high operating rates; this fact was ignored or misunderstood by policymakers who instead interpreted the postwar growth of the industry as the "manifest destiny" of a resource rich country. Excessive capacity additions were brought onstream in the 1970s.

The dangers inherent in such a reversal of course were exacerbated by the stagnation of much of the Canadian economy following the energy crisis. Domestic market growth, which had buttressed the prosperity of the steel industry, disappeared, as can be seen in Table 8–3. One threatening consequence of this was an increased reliance on exports, especially to the United States, in order to maintain operating rates. The increased competitiveness of the Canadian steel industry had led it to continually broaden the definition of its natural market until it finally included parts of the United States. Canadian exports to the United States had always been relatively high, but these had traditionally involved large amounts of semifinished and thus less objectionable products, which were frequently reimported back to Canada for final finishing.[14] As the decade wore on, however, exports of finished products to the United States gradually increased, even though this was a development that most Canadian steelmakers had wished to avoid.

For the Canadian economy as a whole, deteriorating performance virtually destroyed the consensus mode of policymaking in the late 1970s. Programs such as the National Energy Policy (combining Canadian ownership and frontier oil development) virtually gutted the industries they were designed to foster, leading to a massive exodus of capital. The declining value of the Canadian dollar has postponed the full effects of this disastrous shift on Canadian industry. Present exchange rates have maintained and even increased the competitiveness of the Canadian steel industry, for instance; but this masks the fact that the industry's prospects are now much gloomier than they had been only five years ago.

The collapse of the policy-strategy consensus has greatly increased the degree of uncertainty for Canadian steel firms. More specifically, it has eliminated at least two of the three pillars—import policy, tax policy, and high operating rates—on which the specialized, prosperous Canadian industry was built. The need to fund increased federal deficits has led to the reduction of tax incentives; the resource depletion allowances that benefited Canadian steel firms were altered substantially—casualties of the pressure to raise revenues. Furthermore, governmental policy has become indifferent to the necessity of high operating rates—a potentially disastrous change of course. Recent expansion, sometimes with support from provincial governments, has left the Canadian steel industry with significant excess capacity in several product lines, such as hot-rolled sheet.

While governmental petulance certainly played a role in the abandonment of the commitment to high operating rates, the final decision to expand capacity was made by the companies. Recent developments suggest that Canadian firms may have ultimately succumbed to the "American disease" of the 1950s, that is, overoptimism in regard to potential market growth and the permanence of existing comparative advantages. Similar symptoms have been evident in Japan. The likelihood of substantial excess capacity will be a burden to companies that are accustomed to high operating rates and that need such rates of capacity utilization in order to be competitive.

This problem will be exacerbated by another negative development for Canadian steel producers: increasing labor costs. The profitability of the industry through the 1960s and 1970s created pressure for wage increases, but these were generally restrained by the imposition of wage and price controls. In 1981, labor's demand for "catch-up" wage increases led to a long strike, which ended with a 51 percent increase in compensation over the three-year life of the contract. Such settlements are one element in the Canadian economic mosaic of recent years and one cause of the weakness of the Canadian dollar. The overexpansion of Canadian capacity combined with rapidly rising employment costs will appreciably reduce Canadian competitiveness in the long run, unless the Canadian dollar falls continuously—a dubious kind of industrial policy.

Thus, despite its postwar success, the future prospects of the Canadian steel industry now seem clouded—a reversal that can be at least partially ascribed to the collapse of the policy-strategy consensus. It would be incorrect to attribute this solely to the emergence of less supportive government actions. Successful industrial policy in Canada represented a fusion of interests between industry and the state, leading to the implementation of a highly effective industry strategy. While the Canadian government now seems to have ceased striving for consensus, Canadian steel firms seem to have abandoned the strategy that served them so well through the mid-1970s. The collapse of the policy/strategy consensus has left both parties adrift.

CONCLUSION

The interplay between governmental policy and managerial strategy in Canada offers valuable lessons—both positive (from 1955 to 1975) and negative (the late 1970s and 1980s). The Japanese experience offers much the same lesson, at least on a general level. In spite of the vast differences between these countries, their practice of successful industrial policy appear to involve the same fundamentals. This indicates that the formal apparatus (whether planning office, MITI, traditional governmental department, etc.) within which industrial policy is allegedly devised is of only superficial importance. Much more fundamental is the consensus developed between an industry and the government, whatever the formal arrangement by which such a consensus is forged. Successful industrial policies are based upon an industry-government consensus that realistically identifies the existing obstacles to superior economic performance and then establishes a policy environment in which these obstacles can be overcome by appropriate managerial strategies. Effective industrial policy is not formulated by the government; rather, it is constructed in the often murky interaction between industry and the state. There are no formal rules or institutions that can guarantee successful industrial policies; by the same token, there are no rules or institutions that are infallible proof against the corrosion of the consensus that lies at the heart of successful industrial policy.

The fragility of the policy/strategy consensus is both reduced and enhanced by success. Certainly there is little likelihood that a consensus will be forged unless there are also prospects for growth, even though the prospects for growth are increased by the development of successful industrial policies. Both the Canadian and Japanese steel industries faced the challenge of realizing and managing growth opportunities; shaping industrial policies is relatively easy under such circumstances. It is much less clear that these countries will succeed in developing policies to manage retrenchment, the more relevant issue for the United States. Success brings its own problems, however, since successful industries are often targets for governmental policies that are not designed to foster growth.

The prevalence of planning in popular accounts of industrial policy is largely a myth, at least if planning is interpreted to mean overall governmental controls (whether rigid or lax) on the structure and tempo of growth. Whatever "planning" is done in Japan and Canada (or, for that matter, in other Western countries) is less the province of government than of industry. The role of government is actually quite limited: the rather clear-cut identification of strategic sectors, the elimination of potential bottlenecks, and the assurance that the overall goal of increased competitiveness is paramount. In regard to the latter point, it is crucial that the government find some means of ensuring that industrial policy does not degenerate into the state-sanctioned maintenance of an inefficient status quo, which has been the problem with Britain's attempts at planning.

Successful industrial policies are rarely based on a clever identification and exploitation of comparative advantages. In the Canadian case, where an alleged comparative advantage (abundant raw materials) was a more or less explicit element in industrial policy, the perceived advantage actually proves to be nonexistent, at least for steel production. Instead, successful industrial policy views the structure of comparative advantage not as an input but as an output; that is, one of the goals of industrial policy is to turn competitive disadvantages into advantages. This is particularly evident in the policies that fostered the growth of the Japanese steel industry; in the 1950s the application of a comparative advantage rule of thumb would likely have encouraged the Japanese to abandon steel production.

The identification of successful industrial policy with the development of a progressive consensus between industry and the state highlights the difficulty of implementing such a policy approach in the

United States. The conception of industrial policy prevalent in the United States envisions a much more directive role for the government than appears to be the case in the countries we have discussed. "What's good for General Motors is good for the country" seems an unlikely foundation for the reshaping of U.S. economic policy. Yet the perspective it represents is a crude but fair description of the principles applied in the formulation of industrial policy by our more successful competitors. If this perspective is judged on a nonideological basis, the logic behind it appears simple and straightforward. So long as the private ownership of capital is the principal upon which the U.S. economy is based, economic growth and the prosperity it entails depend upon the growth and prosperity of privately held firms. That, quite simply, is the way the system works—or doesn't work.

Notes

1. Lewis Carroll, *Alice in Wonderland* (London: Oxford University Press, 1971), p. 84.
2. See, for example, Douglas Fraser et al., "Crossroads for American Industry" (Washington, D.C.: Industrial Union Department, AFL-CIO, 1982).
3. J. Dresser, T. Hout, and W. Rapp, "The Competitive Development of the Japanese Steel Industry" in Jerome B. Cohen, ed., *Pacific Partnership: United States–Japan Trade* (Lexington, Mass.: Lexington Books, 1972), p. 203.
4. Kiyoshi Kawahito, *The Japanese Steel Industry* (New York: Praeger, 1972), p. 9.
5. Ibid., p. 11.
6. Ibid., p. 26.
7. Ibid., p. 41.
8. Keichi Oshima, "Research and Development and Economic Growth in Japan" in B. R. Williams, ed., *Science and Technology in Economic Growth* (New York: John Wiley & Sons, 1973), p. 313.
9. Ira Magaziner and Thomas Hout, *Japanese Industrial Policy* (London: Policy Studies Institute, 1980), p. 46.
10. Ibid., p. 65.
11. See, for example, Jacques Singer, *Trade Liberalization and the Canadian Steel Industry* (Toronto: University of Toronto Press, 1969).
12. For example, Costeel International, affiliated with Ferrco, owns or is a major partner in two of the most successful U.S. market mills: Raritan and Chapparral. Ivaco owns Atlantic Steel and is a major shareholder in Laclede Steel.

13. Prices and Incomes Commission, *Steel and Inflation* (Ottawa, 1970).
14. Canadian firms traditionally trade extensively in semifinished outputs in order to overcome bottlenecks and thus run at higher operating rates. This practice is rare in the United States.

9 THE POLICY ENVIRONMENT IN THE UNITED STATES

I believe we must come to enforced publicity and governmental control, even as to price, and so far as I am concerned, speaking for our company, so far as I have the right, I would be very glad if we had some place where we could go, to a responsible governmental authority, and say to them, "Here are our facts and figures, here is our property, here our cost of production; now you tell us what we have the right to do and what prices we have the right to charge."[1]

—Judge Elbert H. Gary,
Founder and Chairman,
U.S. Steel Corporation

The American steel industry has lacked the supportive policy environment that benefited the Japanese and Canadian steel industries during the postwar period. Instead, relations between the government and the steel industry have generally been wary—if not hostile. Nevertheless, the negative atmosphere and disparate results should not obscure the ways in which U.S. steel policy conforms to the pattern found in Japan or Canada. For it is the similarities rather than the differences that are the most illuminating aspect of this comparison.

As in Japan and Canada, American steel policies have been based on a perception of steel as a strategic industry. This entailed governmental support in Japan and Canada, where industrial development and international competitiveness were the principal policy goals. Steel's strategic importance has had the opposite effect in the United States, where policymakers have by and large treated the industry as a quasi-public trust. The availability of attractively priced steel for the rest of the economy has been the traditional concern of U.S. steel policy, either to avoid supply disruptions or to suppress the potential inflationary impact of price increases for such a vital industrial input. A similar emphasis was apparent in Japanese and Canadian steel policies. In those cases, however, this goal was sought indirectly, by ensuring that the steel in-

dustry would be an efficient, profitable, and low-cost producer, able to attract investment. In the United States, the goal of maintaining adequate supplies of low-priced steel was sought by cruder means. Political pressure was frequently applied to reduce steel prices, while less attention was paid to economic performance—whether in terms of costs, profits, or other measures.[2]

It may not be surprising that the U.S. government has tended to view the steel industry as a semipublic institution, since such an attitude is prevalent in the great majority of countries that are significant steel producers. What is surprising is that this view has been shared by American steel producers, as is reflected in the quotation that opened this chapter. This convergence of opinion between the government and the industry defines a more important similarity between U.S. steel policy and the experiences described in Chapter 8. As in Japan and Canada, U.S. steel policy has been shaped by the development of a consensus between industry and the state, albeit one of a quite different type. Whereas the strategy-policy consensus has been constructive and forward looking in Japan and Canada, in the U.S. it has been reactionary and defensive. The American steel industry and the U.S. government have traditionally maintained an uneasy and unproductive alliance, the keystone of which is the preservation of a status quo in which both the government and the industry have a stake. Nevertheless, this status quo has ambiguous implications for the expressed goal of each party: adequate supplies of low-priced steel in the government's case and adequate profitability in the industry's.

For historical reasons, the strategy-policy consensus in the United States has been built around the issue of the industry's structure and, more specifically, the role and status of "Big Steel"—the U.S. Steel Corporation. Despite a long history of investigations and threats, the government has traditionally tolerated the concentrated structure of the industry and the leadership of U.S. Steel. The price of this tolerance has been the government's frequent efforts to determine the behavior of the industry as an element of national policy. The industry, led by U.S. Steel, has begrudingly accepted such governmental interference so long as it left the industry's oligopolistic structure intact. Other integrated firms have perceived that they can outperform U.S. Steel and can therefore prosper so long as they remain under the U.S. Steel umbrella and so long as U.S. Steel tolerates the gradual erosion of its market share. As a result of this convergence of interests, the business decisions of the industry—on prices, labor contracts, investment, etc.—have frequently been transformed into political rather than economic issues. Perversely, the politicization of the industry's conduct and performance has accelerated as its economic competitiveness has deteriorated.

The principal goal of the government-industry consensus is not the constructive one of improving industry performance (lowering costs and raising profits) but the reactionary one of maintaining an industry structure that both legitimizes governmental interference and cushions the decline of the traditionally dominant firms. Governmental officials have been prone to suspect steel firms of abusing their market power, particularly in terms of prices, at the expense of overall economic prosperity. This has provided a justification for governmental intervention into the decisionmaking of steel firms. At the same time, however, policymakers failed to show concern over the emergence of the industry's financial difficulties in the late 1950s and early 1960s, in spite of steel firms' persistent amplification of this theme. For many in government, this problem reflected the competitive weaknesses of an oligopolistic industry and was decidedly undeserving of governmental intervention. For the industry, governmental indifference represents a betrayal of the consensus and even of the national interest.

Nevertheless, both parties have generally adhered to the consensus that politicizes the performance of the steel industry, despite the fact that this is becoming untenable. The deteriorating position of U.S. Steel, together with new competition from offshore suppliers and minimills, is fast dissolving the market structure around which the consensus is built. This alone ensures that future steel policy will differ from its traditional pattern.

The reactionary consensus between the steel industry and the American government defines the milieu within which specific steel policies should be judged. Rather than add to the already more than adequate number of studies that detail the governmental policies that have targeted the steel industry, this chapter will attempt to interpret this historical record in terms of the more general themes that were raised above.[3]

STEEL POLICIES

For the first half of the postwar period, U.S. steel policy was primarily concerned with the industry's strength, particularly insofar as its oligopolistic structure enabled it to at least partially circumvent the operation of market forces. Steel was an absolutely crucial input for the entire manufacturing sector, so that its availability at a reasonable price was essential to the maintenance of peacetime economic growth. Furthermore, steel seemed to have a considerable impact on the overall pattern of industrial relations. Steel contracts did then and still do set the pattern for many industries. For that reason, and because increases in steel prices were alleged to generate cost-push inflationary pressures for many industries, the trends in steel wages and prices were viewed as having particular macroeconomic relevance.

The Capacity Debate

In the immediate postwar period, labor relations in the steel industry were a key governmental concern. By the late 1940s, however, this theme had been somewhat eclipsed by questions about the adequacy of American steel capacity. Economists in several governmental departments, some of which had little direct connection to the steel industry,[4] developed projections of potential steel demand in a full-employment economy. According to these estimates, capacity was inadequate, so that steel supplies would prove to be a bottleneck in a rapidly growing economy. Steel shortages would presumably lead to premium prices, generating overall inflationary pressures and potentially limiting business activity in the economy as a whole. Won over by these arguments, and accustomed to the command economy characteristic of World War II, President Truman hinted in his 1949 State of the Union address that the government should consider building steel plants on its own if the industry failed to expand capacity. While the industry protested that by historical standards steel capacity was quite sufficient, these arguments were ascribed to oligopolistic resistance to the lower prices that adequate supplies would ensure. In the late 1940s and early 1950s, a sharp debate was carried out in the press and before congressional committees concerning the adequacy of steel capacity.

From the industry's point of view, the government's concern was asymmetric. The negative repercussions of tight steel markets were given great weight, while the potential effects of excess capacity on the profitability of steel firms were ignored.[5] Subsequent events showed that the industry's perception of capacity needs was more accurate than the government's; as was discussed in Chapter 2, the massive capacity expansion of the 1950s led to deteriorating operating rates and reduced profitability. Nevertheless, it would be fair to say that the government won the dispute, although the victory was of dubious benefit to the national economy. Steel capacity expanded rapidly soon after the issue was joined.

This outcome was partially due to the impact of the Korean War, a period when steel markets were very tight. In 1951, production exceeded 100 percent of rated capacity. The culmination of the capacity dispute occurred in 1952, when President Truman announced the nationalization of the country's steel capacity to forestall the effects of a steel strike on the war effort. This unprecedented step illustrates the bitterness that then characterized relations between the federal government and the industry. Within a few weeks, the Supreme Court had ruled the seizure unconstitutional; the industry's behavior appeared vindicated—in terms of both its expansion efforts and its willingness to accept a strike over contract issues. With the accession of the Eisenhower administration, the momentum of the New Deal and its Fair Deal progeny seemed to have abated, and with them the pressure on the steel industry.

Steel Prices

While the consequences of the capacity dispute—i.e., the overaggressive expansion of the 1950s—persisted well into the 1960s, the issue itself faded rather quickly. Governmental concern with steel prices was more persistent. Until the late 1960s, when the import question began to dominate the government's dealings with the industry, the link between wages and prices in the steel industry and the overall price level was the major theme of U.S. steel policy. In the immediate postwar period, inflation was a principal concern of policymakers, and steel was viewed as one of the levers by which inflationary pressures could be controlled. In fact, the capacity dispute can be viewed in terms of this effort, since what was at stake was less the physical availability of steel than the price at which it would be sold. Once adequate capacity seemed assured, the government's attention shifted to its more traditional concern, the industry's pricing behavior itself.

In the context of the immediate postwar period, steel was the principal locus of the struggle over the pace at which wartime wage and price controls would be loosened. As early as 1945, the government was involved in attempting to mediate between the union and the industry, which had expressed a willingness to grant a wage increase only if a comparable price increase were permitted. In effect, the government made the significant concessions, granting the price rise that made the wage increase palatable to the steel firms. After this, the inflationary pressures that had accumulated during the war could not be maintained, and steel prices led the inflationary march (see Table 2–8).

Governmental involvement characterized every steel contract until at least the late 1960s. Steel agreements always involved three parties: the union, the companies, and the state. The government's stake in these negotiations revolved around the potential inflationary impact of the steel contract, both directly, in terms of steel prices, and indirectly, in terms of steel's role (with auto) as the pattern for wage settlements in the economy as a whole. Government pressure was generally applied for a rapid settlement, and increases in employee benefits were generally viewed more favorably than increases in steel prices.

Surprisingly, this was most clearly displayed by the Eisenhower administration. In 1956, contract negotiations failed to produce an agreement, and a strike was called. Eisenhower's secretaries of labor and the treasury intervened, convincing the industry to agree to increase benefits by almost 25 percent over the three-year life of the contract. Combined with the effects of the 1958 recession, this greatly contributed to the deteriorating profitability of steel producers. Once the 1956 contract took effect, the persistent erosion in the profitability of the industry had begun, and this set the stage for the bitter 116-day strike of 1959.

Until the 1958 recession, however, both wages and prices marched upward together. In 1957, the persistent rise in steel prices provoked another form of governmental intervention: the Kefauver antitrust hearings. The oligopolistic structure of the steel industry had always provoked suspicion on the part of the government, from the earliest antitrust suit against U.S. Steel (the period of the "Gary dinners") to investigations of the basing-point price system in the 1920s.[6] In the late 1950s, the issue was the same, although the vocabulary had shifted. The Kefauver Committee focused attention on the prevalence of administered pricing within the steel industry, reflecting economists' contemporary infatuation with that theme.[7] On the basis of the administered pricing hypothesis, the efforts of steel firms to raise prices in the mid-1950s were castigated as the perverse product of excess market power. Paradoxically, declining operating rates were cited as evidence that steel price increases were unjustified only a few years after governmental pressure had encouraged or forced the expansion of capacity.

As was the case in the capacity dispute, the steel industry's response to the charge of administered pricing was to point to its declining profitability—in essence an admission that steel prices were in fact administered, although a denial that this was a bad thing. "A fair profit" was the industry's consistent justification for its pricing behavior, and it never missed the opportunity to proselytize concerning the role of profits in a market economy. Befitting its status as an oligopoly, however, its recipe for improving profits concentrated on raising prices rather than reducing costs. Since its pricing decisions had become so politicized, the industry learned to pursue its objectives in Washington as much as in the marketplace. A modus vivendi gradually developed that involved tacit union support for steel price increases as the cost of an improved contract; this support could be withdrawn when relations turned bitter, as in 1956. As a contemporary observer commented: "although resort to strikes became common, the outcome of these sometimes bitter struggles was curiously akin to what might have been achieved by overt collaboration between the industry and labor."[8]

The pattern of wage and price increases in the steel industry slowed after the 1959 strike. Nevertheless, the most celebrated confrontation between the industry and the state over prices occurred in 1962. In April of that year, President Kennedy and Roger Blough, then chairman of the U.S. Steel Corporation, waged a highly public dispute over the acceptability of an announced price increase. The Kennedy administration had been seeking to reform the pattern of industrial relations in the United States, to reduce its confrontational character, and thus to dampen the inflationary spiral. This was at least the goal of Arthur Goldberg, Kennedy's secretary of labor, who had been a principal advisor to the United Steelworkers during the 1950s. As secretary of labor Goldberg was actively involved in the negotiations for the 1962 steel contract, stressing that the union should limit wage increases to the level of productivity growth. Such wage restraint on the part of the labor movement would then allow companies to maintain their prices. Steel was the first major test of this approach. Accepting the assurances of the administration that all parties would play their assigned roles, the union agreed to a moderate contract well before the preceding one expired.

Less than a week later, this apparent success was destroyed when U.S. Steel announced an across-the-board price increase. It seemed to most parties that the corporation had thereby reneged on its tacit agreement that a moderate contract would not be followed by a price increase. At the very least, U.S. Steel's action made it appear that the administration could not make good on its commitments to labor—an inference that ensured the failure of any attempt to manage the macroeconomic pattern of wage and price increases. Somewhat obtusely, U.S. Steel pointed to its poor profitability as the justification for its actions. The administration mounted a massive media campaign to reverse U.S. Steel's announcement or to convince other steel producers to break ranks with the traditional price leader. The president was alleged to have characterized businessmen of the Blough ilk as "sons of bitches."[9] Antitrust investigations were threatened, defense procurement was targeted toward firms that failed to follow U.S. Steel, the FBI began investigating potential price-fixing schemes, and so on. Whether because of these efforts or because the market would not support increased steel prices, several companies (most significantly, Inland Steel) announced that they would maintain their prices. Other companies then announced price rollbacks, and three days after its original announcement U.S. Steel followed suit.

This confrontation was the most spectacular incident in the longstanding contention over steel prices; just as importantly, it marked a crucial step toward the collapse of U.S. Steel's status as the industry's price leader. Governmental intervention in steel contract negotiations did not end with the Kennedy–Blough confrontation. It reached its crudest form during the Johnson administration, when Johnson invited union and company representatives to negotiations at the White House and allegedly refused to allow them to use the bathroom until an agreement was reached. In spite of such theatrics, however, steel prices had ceased to command the same attention by the mid-1960s. Whether because of government pressure or because of the entry of new competitors—imports and mini-mills—steel prices rose more slowly than did other wholesale prices through most of the 1960s (see Table 2–8). Prices were restrained, however, only at the expense of profits, which were lower in the early 1960s than at any other time between 1945 and 1980.

Since the early 1970s, the link between steel prices and inflation has rarely been invoked. The Nixon price controls, which were in effect during the steel boom of 1973, did reduce the industry's profits by about one billion dollars according to the government's own investigations.[10] But this program was economywide, and the damage done to steel was inadvertent. By the Nixon presidency, in fact, the problem with steel was the industry's competitive weakness rather than its strength. This was highlighted most starkly by increased import penetration, the course of which was described in Chapter 3.

The Import Question

The industry's response to the import problem has been heavily weighted toward appeals for government relief; insofar as the solution to this problem is defined in terms of extraeconomic controls, it obviously demands governmental action. Efforts to boost the competitiveness of the industry have been pursued with less fervor, possibly because the industry believed its own claims that foreign producers were uniformly less efficient—at least until recently. During the 1960s, the attempt to link the loss of competitiveness with public policy may well have been appropriate. The U.S. industry was then the most efficient in the world, and it did not face the cost disadvantages with which it must now cope. The high value of the dollar, partially a political question under a regime of fixed exchange rates, was one source of the import problem during that decade. Furthermore, the steel import problem is itself merely a reflection of the gradual internationalization of the U.S. economy—a trend that must be linked to the governmental policy of lowering world tariff barriers through the General Agreements on Tariffs and Trade (GATT). While these broader issues have rarely entered the debate, the industry's concentration on the import issue has ensured that trade has been the chief theme of steel policy since the mid-1960s.

Almost all the potential permutations of trade policy, from benign neglect to quotas, have been applied to the steel import problem. The first explicit intervention by the U.S. government was in the late 1960s, when imports had surged to almost 17 percent of the U.S. market. The Nixon administration then negotiated self-administered commitments from major steel exporting countries that they would limit their exports to the United States. The chief target of this program was Japan. The commitments themselves, in effect a quota system, were euphemistically described as "Voluntary Restraint Agreements" (VRAs).

The fatal weakness of this program was that it limited the aggregate tonnage exported by the countries involved, so that many exporters simply shifted to higher value products, maintaining or even increasing the dollar value of their U.S. sales. Imports from nonsignatory countries increased, and there were disputes about whether the bargain was kept. At any rate, it was a strictly voluntary agreement, one that contained no enforcement mechanism. In most years, its overall results were meager. Imports did slow gradually, but it is unclear whether this can be ascribed to the agreement. Most importantly, the arrangement did not lead to any increase in U.S. investment or any reversal of the industry's competitive decline. Since this was the underlying purpose of the VRAs, they must be judged a failure.[11]

In spite of this, the original agreement, which had a life of three years, was renewed once. Following this extension, it was allowed to die a natural death. By 1973, the world steel boom and U.S. price controls had drawn imports away from the United States, so that the problem had faded away. Imports then seemed to have reached a plateau of 15 percent of U.S. consumption, and the industry optimistically assumed that this status quo would be maintained.

This optimism was unjustified. As world steel demand fell after the boom years of 1973–74, the import share of the U.S. steel market again rose. The industry's response was to pursue the legal remedy of antidumping suits.[12] The Carter administration countered with the Solomon Plan, the first attempt to develop a comprehensive set of policies to reverse the declining prospects of domestic steel producers.[13] This was an implicit recognition of the seriousness of the problem, and several agencies undertook thorough analyses of the industry's problems at about the same time.[14] The Solomon Plan suggested policy changes in several areas: tax policies, antitrust, environmental regulations, R&D support, and so on. Its centerpiece, however, was a trade program that was designed to obviate the industry's reliance on court actions. This program was the trigger price mechanism (TPM).

The TPM was a stopgap measure designed to placate the U.S. industry without resorting to quotas. With the help of the Japanese industry, the U.S. government developed a model of Japanese steel costs, which were assumed to be the lowest in the world. Steel entering the U.S. market at prices below Japanese costs, including a provision for an 8 percent profit, were supposedly subject to an immediate dumping investigation. The constructed Japanese price thus acted as a minimum price for imported steel; imports at prices below this level would trigger an accelerated and automatic dumping investigation. Although the industry was not enjoined from initiating dumping complaints under the trigger price system, the assumption was that this would be unnecessary. In fact, the government stressed that the filing of suits would mean the termination of the TPM.

Beneath the surface, it appears that the true intent of the TPM was to stabilize the import share of the U.S. market at roughly 15 percent. The program implicitly assumed that U.S. firms, which consistently claimed to be the low-cost suppliers to the American market, would be able to compete and to earn a fair return if import prices reflected Japanese costs plus transport to the United States. Japanese producers were quick to pick up on this "understanding" and unilaterally limited their U.S. exports to six million tons a year. Other producers failed to accomodate this understanding. European firms, for instance, found the TPM a boon. The trigger price at which their output was sold was well below their costs, so that the program in effect sanctioned dumping by European producers. This was deemed irrelevant, however, given the implicit assumption of U.S. competitiveness at the trigger price. Unfortunately, either the American industry overestimated its competitiveness or the data submitted to the U.S. government understated Japanese costs. At any rate, the TPM failed to preserve the domestic industry's market share and provide a "fair" return. In a sense, European suppliers never "understood" the true intent of the program and quickly moved to fill the gap left by Japanese self-restraint. These weaknesses were compounded by the fact that enforcement of the program was lax—perhaps inevitably so, given the system's ungainliness and the traditional distaste of the Treasury Department, which monitored the program, for any interference with free trade.

The principal consequence of the TPM was to establish a somewhat porous price floor for the U.S. steel market. This was the clearest sign that the government's stance toward the industry had shifted since the early 1960s, when it sought to suppress steel prices. In the context of the world steel crisis which gained momentum after 1975, the TPM ensured that prices in the American market would exceed those prevailing in world markets by an increasingly significant margin. This threatened the international competitiveness of industries that used American steel, and it made the U.S. steel market the most attractive in the world, providing a great incentive for steel producers to export to the United States.

As was the case with the VRAs, the U.S. government itself judged the TPM a failure.[15] By the late 1970s American producers were persistently grumbling about the program, and in 1980 U.S. Steel initiated legal actions against a broad range of exporters. This led to the first termination of the TPM, although it was strengthened and reinstated in the waning days of the Carter administration, when U.S. Steel withdrew its suits. While there was little enthusiasm for it in government, it was still viewed as the best alternative and had become an acceptable component of relations with our major trading partners.

Partially as a result of the increasingly overvalued dollar, steel imports surged again in 1981, when the U.S. steel market was relatively weak. In response, domestic producers prepared a massive legal assault, calling for both countervailing duties (in response to foreign subsidies) and dumping penalties. When the suits were filed in early 1982, the trigger price program, which by then no longer had a domestic constituency, was dropped.

Unlike the VRAs and the TPM, the legal actions of 1982 targeted European producers, although some other countries (e.g., Brazil and South Africa) were also involved. Despite the overvalued dollar, steel producers in several European countries—especially France, Italy, Belgium, and the U.K.—would be uncompetitive in most U.S. markets were it not for significant governmental subsidies. In the summer of 1982, the U.S. Department of Commerce found in favor of the U.S. industry in many of the countervailing duty cases. This provoked drawn-out negotiations among the U.S. government, U.S. steel producers, and the Common Market authorities. Although the actual imposition of duties where subsidies were found would have addressed the actual market distortions that are prolonging the world steel crisis, political considerations always made this the most unlikely result. At the last hour, the parties agreed to an enforceable system of quotas on import tonnages from the European Economic Community (EEC) as a whole. Unlike the VRAs, these quotas had legal standing and covered specific product categories.

Oddly enough, the European market share defined by this quota system was not much below the traditional level. This result suggests that American producers were less interested in eliminating unfair competition than in reaching an accord that would sustain an adequate price level. Since European producers (both subsidized and unsubsidized) are guaranteed a specific share of the market, they have less incentive to cut prices in order to increase the quantity sold. Thus, even the most recent trade program reflects the industry's traditional concern that prices be adequate to provide a satisfactory rate of return.

Having shored up their position on the European front by the end of 1982, U.S. producers turned their attention to the Far East. In early 1983, U.S. Steel et al. filed suit against Japanese producers on the basis that they had agreed to drastically reduce their exports to Europe but had failed to offer the same benefit to U.S. producers—allegedly a violation of the GATT treaty. Although this suit was rejected by the Reagan administration, it reflects the industry's efforts to construct an effective limit on imports into the U.S. market and thus to protect the domestic price level. The U.S. government is being drawn into this effort, as foreign governments have been drawn into the competitive difficulties of their steel industries.

The persistent resort to trade barriers is the final theme in the dénouement of the industry-government consensus that has shaped American steel policy throughout this century. Whereas steel policy had traditionally focused on restraining steel prices, the government is now implementing policies that indirectly generate higher steel prices. While designed to boost the industry's profitability, such policies are just as likely to attract further entry on the part of mini-mills and foreign suppliers not covered by import restraints. Trade policies are powerless to offset the forces that have determined the competitive decline of integrated steel production in the United States, and thus they are powerless to maintain the status quo around which the consensus is built.

MACROECONOMIC POLICIES

Since at least the late 1960s, the U.S. steel industry has complained that macroeconomic policies have undermined its competitive prospects and, by implication, those of other basic manufacturing industries. While the industry's focus has typically been on macroeconomic programs that directly affect cash flow (e.g., taxes), there are numerous ways in which macro policies have altered its standing in the market. Any governmental policy has sectoral effects, so that the policy environment provided by macroeconomic programs has conditioned the steel industry's prospects. This is a theme which extends far beyond the scope of this book, however—especially since macroeconomic policies in the United States have not been designed to benefit specific sectors. Thus our discussion will necessarily be brief. We have already obliquely described the extent to which incomes policies—direct governmental efforts to determine wages and prices—have affected the steel industry. The following comments will therefore be devoted to other macroeconomic policies.

Monetary Policy

The steel industry has devoted much less attention to monetary policy than to other macroeconomic policies of the federal government, probably because the effects of such policies on the industry's cash flow are less direct. Nevertheless, monetary policy has shaped the competitive prospects of the steel industry in at least three ways: through inflation, through interest rates, and through exchange rates.

Inflation. Overall price trends should be viewed as the responsibility of the state—even if the specific policies that cause the inflationary disease are subject to debate, as are the policies that would constitute a nonrecessionary antidote. Governmental tolerance or encouragement of inflation has significant sectoral effects. Highly concentrated industries with strong unions (such as steel and autos) are especially vulnerable to the corrosive effects of inflation, since their employees are more likely to be able to maintain, or even increase, the value of their real wages, whereas this has proved difficult for workers in other industries since the late 1960s. This tends to open a gap between employee compensation in such industries and the manufacturing average, which increases their vulnerability to foreign competition (see Chapter 3). Even if one assumes that the costs and prices of all industries increase at the same rate, however, persistent inflation alters the relative prospects of different industries due to differences in the physical or technological features of production processes.

Inflation affects the valuation of assets and the attractiveness of investments. Since price increases boost the cost of replacing facilities, they indirectly tend to preserve the economic viability of existing equipment. This reduces the incentive to invest. The more capital intensive an industry, the greater the inflationary bias against new investment. Furthermore, the economic uncertainty and high nominal interest rates characteristic of an inflationary environment are particularly threatening to long-term investments. Hence, the inflationary bias against new investment is particularly pronounced for a capital-intensive industry like steel, the assets of which tend to be long lived.

Insofar as persistent inflation undermines the viability of such capital-intensive industries, it encourages an increase in the relative importance of the so-called service sector, which tends to be relatively labor intensive. Productivity growth tends to be slowest in service industries.

At the same time, retarded investment in capital-intensive industries limits their productivity growth. The combination of these factors exacerbates the problem of stagflation, contributing to a vicious cycle that policymakers have failed to break.

Interest Rates. High rates of inflation inevitably wreak havoc with the social and individual structure of time preference, which is the underlying determinant of interest rates. Whereas inflation has weakened the relative position of capital-intensive industries like steel in the U.S. economy, this has been partially offset by the failure of real interest rates to keep pace with the rate of inflation. In real terms, long-term interest rates in the U.S. economy have been extremely low throughout the postwar period; from 1973 to 1980 both long-term and short-term rates have been negative in terms of their real return.[16] Such a situation has obviously made it relatively attractive to borrow money, and this is one reason that the indebtedness of the steel industry (as well as other sectors of the U.S. economy) grew so dramatically during the postwar period. The exceedingly low level of real interest rates has presumably made it easier for the industry to invest during a time when most other factors made it more difficult.

The high real interest rates of the early 1980s have redressed the balance and have made it extremely difficult for the steel industry to raise funds for investment. As a result, the industry has increasingly turned to more novel forms of financing.[17] If interest rates remain high by historical standards, this is likely to accelerate the structural transformation of the U.S. economy away from heavy industries like steel. The profitability of the industry is so low that it must rely on the combination of high inflation and low real interest rates to prop up its attractiveness as an outlet for investment. Without these props, and in an environment where capital is priced at its real opportunity cost, the unprofitable character of many steel industry investments will be much more apparent.

Exchange Rates. The development of a truly international steel market has been one of the fundamental changes confronting steel firms during the postwar period. One of the principal consequences of this trend is that cost competitiveness is now more subject to factors that are beyond the control of the firm. As in the past, a firm's performance is determined by the efficiency with which inputs are used and the price at which they are purchased. Within an international context, however, competitiveness is also determined by the structure of exchange rates. Physical relationships of productivity and even cost relationships in the production process can be altered dramatically by movements in exchange rates—a possibility that became painfully apparent to American manufacturers in 1982.

It could be argued that exchange rates are not really determined by governmental policy at the present time, and there is at least some truth in that claim. In a fixed exchange rate regime, however, currency values are partially determined by government policy. The international monetary system was characterized by such a fixed-rate regime from the end of World War II until the the system began to be dismantled by the dollar devaluations of the early 1970s. Before then, the exchange rate of the dollar had shifted only as the gold values of other currencies (particularly the German mark) were altered. This system of relatively fixed parities, based on the dollar, was established at a time when the productivity advantage of the U.S. economy was extreme; the dollar's value was set accordingly. By the late 1960s, the relative strength of the U.S. economy had eroded considerably, and this trend was exacerbated by the maintenance of the parities established over twenty years before.

The increasingly overvalued dollar in effect subsidized imports into the U.S. market (alternatively, taxed U.S. exports). Had exchange rates been flexible, adjustment would have weakened these effects, providing some relief to domestic manufacturers. The increasingly overvalued dollar was one reason U.S. steelmakers were unable to slow the penetration of their market by foreign suppliers during the 1960s. Despite substantial investments, greater efficiency in the use of inputs, and employment costs that were declining relative to the manufacturing average, domestic producers' share of the domestic market fell from 97.1 percent in 1958 to 83.3 percent in 1968. By the time exchange rate adjustment finally began in 1971, the deterioration of the U.S. industry's position had advanced to the point where this provided little relief.

Other basic industries in the United States also lost competitive ground due to the overvaluation of the dollar in the late 1960s.[18] At the same time, however, certain sectors of U.S. business benefited from (or exploited) this imbalance. Although the overvalued dollar taxed domestic manufacturing, it subsidized financial operations abroad, especially foreign investment. Foreign investment by American steel firms was generally devoted to raw materials extraction; perhaps because steel operations are generally viewed as having strategic significance and are thus subject to political controls, very little international investment has occurred in integrated facilities. Hence, steel was an industry that experienced only the negative effects of an overvalued dollar; its relative decline within the U.S. economy is partially due to this fact. In fact, the exchange rates that prevailed from the end of World War II to the early 1970s could be viewed as one of the factors that has propelled the transition of U.S. economic activity away from manufacturing and toward services, since service industries are much less vulnerable to the foreign competition that an overvalued dollar encourages.

The dollar devaluations of 1971 and 1973 and the reform of the international monetary system (the abolition of fixed exchange rates and the institution of floating rates) finally removed the exchange rate "tax" placed on U.S. manufacturing. The pattern of the 1960s was repeated in the early 1980s, however—this time because of high interest rates. While the government's role in determining exchange rates is much less direct under a floating-rate regime, the link between the high value of the dollar and high interest rates is unmistakable. By the fall of 1982, the real effective exchange rate of the dollar approached predevaluation levels.[19] While the U.S. has maintained a surplus on current account throughout this period, this has been due to a spectacular increase in foreign earnings, largely the result of higher interest on loans by U.S. financial institutions. The trade deficit, by contrast, has reached unprecedented levels, reflecting the extent to which U.S. manufacturers have been devastated by exchange rate fluctuations.

The resurgence of the dollar and the increasing fragility of the international financial system obviously define a very significant issue for economic policymakers. The predominance of financial flows in determining exchange rates is a new development, and the existing system was not designed to cope with it. Such issues are beyond the scope of this book. Nevertheless, it is probably true that the value of the dollar in the late 1960s and in the early 1980s has had a significant effect on the competitiveness of the U.S. steel industry and has contributed to the restructuring of the American economy at the expense of steel and other basic industries.

Fiscal Policy

Governmental policies could be designed to compensate for inflationary biases against long-term investments, particularly through the tax system—e.g., decreasing the tax on long-term investments relative to the tax on short-term investments. Similarly, tax policies could be used to offset the sectoral impact of shifts in exchange rates. Such options have not been pursued. Instead, the U.S. tax system has exacerbated the sectoral bias inherent in inflation. This has been most apparent in the depreciation schedules that prevailed until 1981. By linking depreciation allowances to the estimated physical life of equipment, federal tax policy effectively ignored the link between inflation, tax policy, and investment. Steel, for instance, faced a depreciation period of eighteen years—one of the longest for any industry.[20] In an inflationary period, nominal benefits provided by depreciation represent increasingly smaller real benefits as the depreciation schedule lengthens. More rapid depreciation thus offsets the bias favoring short-term investments.

Even more significantly, U.S. tax policies were not developed or applied according to the prevailing international standards. Until the 1980s, American steel producers faced a depreciation schedule longer than that applied to any other steel industry in the world. This disadvantage was reduced slightly as a result of the Solomon Plan, but even after this reform U.S. producers faced a write-off period of fifteen years, compared to one year in Britain, two years in Canada, and so on.[21] The tax reforms instituted by the Reagan administration, however, have shortened the depreciation schedule drastically, so that U.S. steel producers now face a write-off period that more or less conforms to international standards. Furthermore, these measures eliminated the guideline life concept, so that all U.S. industries now face the same depreciation schedule. Tax leasing provisions were also enacted (although these are politically vulnerable) to permit less profitable industries, like steel, to gain the full benefit of depreciation allowances.

In terms of actual corporate tax rates, there is little variation internationally, so that these do not generally constitute disadvantages for American steel producers. Nevertheless, taxation provides a prime example of how U.S. policies have tended to be developed and applied in an international vacuum. Regardless of the domestic equity effects of a specific piece of tax legislation, tax policies that deviate from international norms necessarily affect the competitive prospects of an industry involved in an internationalized market. Even policies that are scrupulously and equitably enforced on all segments of the economy will thus have sectoral effects determined by the extent to which different sectors are subject to international competition. Manufacturing is far more vulnerable to international competition than is the service sector; thus, U.S. tax policies that have been less generous than those prevalent in competing nations have tended to accelerate the structural transformation of the U.S. economy towards service industries.

Industrial Policies

Even though industrial policy is an underdeveloped component of the total economic policy apparatus, several regulatory agencies pursue policy goals that have direct sectoral implications—even though they are formulated in general terms. Regulatory programs can thus be viewed as industrial policies, supplementing more traditional initiatives such as subsidies and differential tax treatment. Three types of industrial policies have had a significant impact on the postwar U.S. steel industry: antitrust policy, environmental policy, and federal support for research and development (R&D).

Antitrust. The U.S. steel industry has had a long and uneasy relationship with the antitrust authorities of the U.S. government, beginning with the formation of the U.S. Steel Corporation in 1901—the most spectacular consolidation in the history of the U.S. economy. Prior to World War I, the Justice Department filed suit to break up U.S. Steel, but in 1920 the Supreme Court denied this suit on the basis that U.S. Steel had not used its market power to drive its competitors out of business. This landmark decision signaled an end to the activist interpretation of antitrust rules that had led to the breakup of the Standard Oil and American Tobacco Trusts before the war. Furthermore, it sanctioned the deemphasis of price competition in the American steel industry—a legacy that left the industry poorly prepared to cope with entry on the part of foreign suppliers and mini-mills in the 1960s.

In the postwar period, antitrust agencies have pursued several cases against the industry, covering both alleged price fixing (via the medium of the industry's trade association) and proposed mergers. The most significant of these cases was the Justice Department's refusal to allow the proposed merger between Bethlehem Steel and Youngstown Sheet and Tube—an issue discussed in Chapter 4. More recently, the Justice Department's failure to contest the merger between Ling-Temco-Vought (Jones & Laughlin) and Lykes (Youngstown)—in spite of the fact that this involved the industry's fifth and eighth largest firms—may have indicated a more liberal attitude toward steel industry mergers. On the ogher hand, it may not have signaled any change in the interpretation of antitrust. Since Youngstown was at that time clearly at death's door, the merger could have been allowed on the basis of traditional "failing company" principles, according to which the enforcement of antitrust provisions may be relaxed.

In spite of the possibility that recent actions by antitrust authorities (e.g., the abandonment of the IBM case) reflect a revision in the way these regulations will be interpreted and applied in the future, anxieties about antitrust regulations still affect the decisionmaking process within the steel industry. Particularly in a period of retrenchment and reduced cash flow, joint actions could have salutary effects. Efficient facilities at plants owned by different firms could be linked to strengthen overall industry performance. Facilities with significant economies of scale could be built and run jointly; alternatively, firms could agree to specialize in certain operations, thereby reducing the need to maintain expenditures in other areas. There are several obstacles to such programs, so that a relaxation of antitrust principles would not necessarily lead to their adoption. Nevertheless, antitrust considerations are one important reason that steel firms have generally failed to investigate the potential benefits of this sort of collective response to the problems of their industry.

Environmental Regulation. Steelmaking is one of the principal sources of industrial pollution in any country with a large integrated steel industry. In the United States, the steel industry is estimated to have been the source of 10 percent of total particulate emissions and one-third of industrial waste-water discharge.[22] Increased environmental consciousness has led to a worldwide effort to reduce environmental pollution, and this effort has quite understandably had a major impact on steelmakers. Whereas the macroeconomic effects of environmental regulation (e.g., on GNP) are slight, the sectoral effects can be enormous. Because of its status as a major polluter, steel is one of the industrial sectors that will be most affected.

According to industry reports, roughly 20 to 25 percent of total steel industry investment since 1975 has been devoted to environmental expenditures.[23] As was noted in Chapter 6, such investments do not raise the efficiency of the industry (the opposite result is more likely) and thus do not reduce the private costs incurred by the industry—even though they may drastically reduce the social costs of steel production. In an environment characterized by severe capital constraints, such as the 1970s, mandated environmental expenditures can significantly limit an industry's ability to undertake modernizing investments.

Insofar as the international competitors of the U.S. industry have been subject to equally stringent environmental regulations, the net penalty represented by such regulations is reduced or eliminated. Environmental regulations in Japan appear to be at least as strict as those in the United States, and this may also be true of several European countries. Nevertheless, it is probably also true, as the industry claims, that the actual costs of environmental regulation to steel firms are greater in this country than abroad. In the Japanese case, much of the environmental equipment was designed into greenfield plants; this is always cheaper than adding such facilities to an existing plant (retrofitting). In Europe, the enforcement of environmental regulations is alleged to be more flexible; it must be admitted that it would be difficult to imagine a more cumbersome and legalistic system than the one that exists in the United States.

There are solid arguments that could be marshaled in support of the way in which the United States has pursued the goal of environmental improvement. Regulations have defined acceptable levels of pollution, and polluters have in principle been forced to adhere to these standards. The costs of compliance are thus borne by the offending parties; the biggest polluters incur the greatest costs. Yet this approach, regardless of its moral appeal, represents an industrial policy as well as a regulatory policy. Integrated steel production has probably been the sector most affected by environmental regulations, largely because of the availability of competing products from other industries, relatively nonpolluting domestic steel firms (the mini-mills), and offshore suppliers. The sectoral implications of environmental policy could conceivably have been offset by socializing the costs of compliance, that is, by subsidizing environmental expenditures in some way. Since this option was generally not exercised, it should be expected that one of the consequences of environmental regulation will be a contraction in those sectors that are significant polluters, as the social costs of pollution are privatized via the regulatory mechanism.

Support for Research and Development. Governmental subsidization of research and development is in many countries, including the United States, a crucial component of industrial policy. The most highly developed form of industrial policy in the United States concerns the defense industry, especially its high technology segment. Defense industries receive substantial governmental support, either in terms of a guaranteed return on investment, the promotion of export sales, or the subsidization of research expenditures. Several of the most dynamic sectors of the U.S. economy (e.g., semiconductors) have been nurtured by federally subsidized research, and several emerging industries—even some that do not seem to have immediate relevance for national defense (e.g., robotics)—are now receiving such support. Defense expenditures are the U.S. equivalent of Japanese or French support for the high technology sector.

The argument is often made that a healthy steel industry is essential to national defense; even if this were true, it is certainly not reflected in governmental support for steel industry R&D. Less than 2 percent of total steel R&D expenditures are funded by the federal government; in the aerospace industry, by contrast, 72 percent of total research and development expenditures come from the government.[24] Not surprisingly, the U.S. aerospace industry is the world leader technologically—a status that has long since disappeared for the American steel industry. Technological innovations in steel production could contribute to the realization of important national goals (e.g., energy conservation), but this has not led to any significant federal commitment to funding steel-related R&D. In several other countries (particularly in Europe), extensive research in relatively exotic steelmaking techniques is being carried out with government support. Such directly commercial research is less likely to be undertaken by the U.S. government unless it can be camouflaged as defense related. This naturally affects the relative prospects of different sectors within the U.S. economy.

CONCLUSION

The link between government policy and industry performance can be assessed along two lines. The first concerns how overall government policies affect the relative prospects of different industries, and the second concerns the policies that have been implemented in order to affect the performance of a specific industry.

If we look first at the way macroeconomic policies have shaped the performance of the American steel industry, there is a large body of evidence suggesting that these policies have contributed to the industry's decline. Despite this evidence, the sectoral effects of macroeconomic policies are rarely the subject of economic debate. Ideological factors maintain the fiction that government intervention is or should have a neutral sectoral impact. The postwar role of the United States in the world economy ensured that the sectoral effects of macroeconomic policies within the United States were often eclipsed by other concerns—e.g., the maintenance of the international monetary system. Finally, U.S. industrial policy has been banished to the constitutionally ambiguous sphere of the regulatory agencies, and this perpetuates the contradictory, unconscious, and legalistic nature of the policy environment faced by American industries. Such factors camouflage the sectoral consequences of macroeconomic policies in the United States.

Neither U.S. industry nor the U.S. government has paid adequate attention to the implications of the increasing internationalization of the American economy. Since the 1950s, tariff barriers have fallen, increased trade has tended to equalize input prices, and technology transfers have narrowed international productivity differentials. As a result, U.S. manufacturers have become increasingly vulnerable to foreign competition. U.S. policies rarely sought to adjust to this development. The dangers inherent in an overvalued dollar were given short shrift, tax and regulatory policies were not judged by international standards, and so on. In general, the U.S. policy environment favored the relative expansion of the service sector—partially because services are much less vulnerable to foreign competition. Within this milieu, basic manufacturing has experienced a gradual secular decline.

The effects of macroeconomic policy on the steel industry have been compounded by the pattern of steel-specific policies. The industry has complained about governmental hostility for decades, while policymakers have long been suspicious about the potential abuses inherent in the industry's oligopolistic structure. Yet actual steel policies have been based on a destructive consensus between steel firms and the state. Steel producers, for their part, have been unwilling to foresake the traditional pattern of polite rivalry that is characteristic of an oligopolistic industry. The industry's oligopolistic status has in turn legitimized governmental intervention in decisions concerning prices, capacity, and so on. While this consensus sustained an uneasy equilibrium through the first half of this century, it has proved especially destructive over the last two decades. New competition, from foreign suppliers and from domestic mini-mills, has gradually undermined the viability of the traditional integrated oligopoly and with it the traditional consensus between the steel industry and the state.

Unfortunately, the industry has responded to this threat by further politicizing its performance. Far too often, integrated steel firms have sought to maintain competitiveness by efforts in the political sphere rather than in the plants. Lobbying efforts, particularly around foreign trade, have entailed the construction of increasingly fragile coalitions—with, for example, the United Steelworkers of America, the Congressional Steel Caucus, and other industries. This complex web of partially converging interests has actually harmed the industry more than it has helped. Steel firms have been reluctant to cut back unprofitable operations for fear of alienating politicians who might provide some trade or tax relief. Little attention was paid to restraining wages or reforming work rules (until 1982), since an alliance with the union was viewed as offering some political benefits. Most unfortunately, perhaps, the commitment of resources to public relations campaigns and political lobbying has made it more difficult for steel firms to recognize the fundamental economic changes that are eroding both their competitiveness and their political clout.

Notes

1. Testimony of Judge Gary before the Stanley Commission of Congress in 1911. Quoted in Ida M. Tarbell, *The Life of Elbert H. Gary; The Story of Steel* (New York: Appleton, 1925), p. 232.
2. The French steel industry, despite significant financial assistance from the French government, has been subject to even stricter price controls; this is one of the reasons that that industry is relatively inefficient by international standards.

3. For more detailed discussions of governmental steel policy, see Federal Trade Commission, *Staff Report on the United States Steel Industry and Its International Rivals: Trends and Factors Determining International Competitiveness* (Washington, D.C.: U.S. Government Printing Office, 1977); and Government Accounting Office, *New Strategy Required for Aiding Distressed Steel Industry* (Washington, D.C.: U.S. Government Printing Office, 1981).
4. The principal supporter of capacity expansion was Louis Bean of the U.S. Department of Agriculture.
5. One industry spokesman was very explicit about the industry's view:

   With the little band of Keynesian economic planners who have their little "compartments" in every department and bureau of the Government in Washington, the greatest crime of which the steel industry is guilty, is the crime of failing to keep abreast of the "managed" expansion of the economy . . . this writer, for one, has never seen one scintilla of evidence that government planners recognized any basic inconsistency in Washington's attitude toward the problem. They have never, in fact, indicated that they were aware that there was any relationship between prices and profits, on the one hand, and the rate of industrial growth on the other.

   E.H. Collins, quoted in Henry W. Broude, *Steel Decisions and the National Economy* (New Haven, Conn.: Yale University Press, 1963), p. 19.
6. Gary dinners were given by Judge Gary, chairman of the U.S. Steel Corporation, for executives of other steel companies in the first decade after the formation of U.S. Steel. At these dinners, Gary exhorted his competitors to refrain from cut-throat competition and to seek stability in their markets. Pittsburgh-plus (or basing point) pricing was a system that regularized steel prices as though it were all produced and shipped from Pittsburgh—regardless of the specifics of an actual sale. The FTC successfully filed suit against this practice after World War I.
7. Administered pricing refers to the allegation that firms in highly concentrated industries retain absolute profit levels by manipulating the markup on sales, regardless of actual trends in demand. This practice could contribute to an explanation of price stickiness during recessions. The classic reference to the theory of administered pricing is Gardiner C. Means, *Industrial Prices and Their Relative Flexibility*, 74th Congress, 1st sess., Senate Document Number 13 (Washington, D.C.: U.S. Government Printing Office, 1935).

8. Grant McConnell, *Steel and the Presidency–1962* (New York: Norton, 1963), p. 55.
9. In the first flurry of activity after Blough visited the White House, Kennedy allegedly commented: "My father always told me that all businessmen were sons of bitches, but I never believed it until now." See Wallace Carroll, "Steel: A 72-Hour Drama with an All-Star Cast," *New York Times*, April 23, 1962, p. 25. At a subsequent press conference, Kennedy stated: "Now the only thing that was wrong with the statement was that . . . it indicated that he was critical of the business community . . . That's obviously in error, because he was a businessman himself. He was critical of the steel men." See *The Kennedy Presidential Press Conferences* (New York: E.M. Coleman, 1978), p. 288–89.
10. Federal Trade Commission, *Staff Report on the United States Steel Industry*, p. 264. This volume also contains a detailed description of government efforts to control steel prices during the 1960s (pp. 267–305).
11. See Government Accounting Office, "Economic and Foreign Policy Effects of Voluntary Restraint Agreements on Textiles and Steel" (Washington, D.C.: U.S. Government Printing Office, 1974).
12. In 1975, the Ford administration imposed import controls on specialty steels. Since these make up a relatively small portion of the total steel market, we will forgo discussion of this episode.
13. "A Comprehensive Program for the Steel Industry, Report to the President Submitted by Anthony M. Solomon, Chairman, Task Force, December 6, 1977," reprinted in *Administration's Comprehensive Program for the Steel Industry: Hearings Before the Subcommittee on Trade of the Committee on Ways and Means, House of Representatives* (Washington, D.C.: U.S. Government Printing Office, 1978).
14. Federal Trade Commission, *Staff Report on the United States Steel Industry*; and Council on Wage and Price Stability, *Prices and Costs in the United States Steel Industry* (Washington, D.C.: U.S. Government Printing Office, 1977).
15. See Government Accounting Office, "Administration of the Steel Trigger Price Mechanism" (Washington, D.C.: U.S. Government Printing Office, 1980).
16. See Benjamin Friedman, "Postwar Changes in the American Financial Markets" in M. Feldstein, ed., *The American Economy in Transition* (Chicago: the University of Chicago Press, 1980), p. 40.
17. U.S. Steel, for instance, is financing its new tubular mill at its Fairfield Works by means of funds provided by consuming industries

(chiefly oil companies), leasing of equipment, and export incentives from the Italian government (much of the equipment is Italian).

18. Otto Eckstein, *The Great Recession* (Amsterdam: North Holland, 1978), p. 107.
19. Real effective exchange rates are regularly printed in a monthly publication, Morgan Guaranty Trust Company, *World Financial Markets*. The concept of real effective rates is discussed at length in the May 1978 edition.
20. American Iron and Steel Institute (AISI), "Steel at the Crossroads" (Washington, D.C.: AISI, 1980), p. 48.
21. Office of Technology Assessment, *Technology and Steel Industry Competitiveness* (Washington, D.C.: U.S. Government Printing Office, 1980), p. 59.
22. Clifford Russell and William Vaughan, *Steel Production; Processes, Products, and Residuals* (Baltimore: Resources for the Future and Johns Hopkins University Press, 1976), p. 6.
23. Calculated from AISI, *Annual Statistical Report, 1981* (Washington, D.C.: AISI, 1982), pp. 9, 10, and 13a.
24. National Science Foundation, "Research and Development in Industry, 1980" (Washington, D.C.: U.S. Government Printing Office, 1982), pp. 11 and 18.

10 STRATEGIES FOR SURVIVAL TO THE YEAR 2000

Now, here, you see, it takes all the running you can do, to keep in the same place. If you want to get somewhere else, you must run at least twice as fast as that![1]

—*Lewis Carroll*
Through the Looking Glass

Deteriorating international competitiveness, shrinking markets, and lagging profitability make up the economic environment that now confronts the American steel industry. Against this background, its former prosperity seems a distant dream. The once mighty kingdom has been laid low, battered by fundamental socio-economic changes and crippled by its own rigidity and inertia. The industry's altered status reflects broader forces reshaping the entire U.S. economy. Rates of growth in output and productivity have lagged, and many American manufacturers have lost ground in terms of international competitiveness. Rates of inflation, particularly in real energy costs, have increased spectacularly, disrupting capital markets and investment patterns. Far-reaching technical changes have created new industries and altered the standards of efficiency for even the most traditional sectors of the economy. Such developments have made the economic landscape of the 1980s radically different from that prevalent during the first two decades after World War II.

Since the 1950s, the steel industry has lost the central position it once occupied in the U.S. economy. While still an important component of GNP, its market has lagged the overall economy. Domestic steel consumption has risen at an anemic rate since 1950, offering the industry none of the benefits growth provides. The challenges presented by slow growth have been aggravated by the emergence of a truly international steel market, a market dominated by major competitors whose performance has overshadowed that of the American industry. Moreover, the domestic steel market has fragmented during the postwar period, especially since the late 1960s. This trend is associated with more sophisticated technical and quality requirements on the part of steel-consuming industries, with shifts in relative prices, and with changing tastes. Constrained by slow growth in demand, challenged by new competition, and confronted with diverging consumer requirements, U.S. steel producers have been forced into a very different market from the one they knew in the first half of this century.

Significant shifts in relative costs and technology have also gripped the industry. Iron ore costs in the United States have accelerated relative to newer sources (e.g., Australia) and relative to substitute materials (e.g., scrap). Energy and labor costs have risen tremendously in the 1970s, and the productivity advantage that had protected the industry from the potential consequences of high employment costs has been eroded. Capital costs have become exorbitant for integrated producers, fostering the development of alternative capital-saving techniques, which are lowering the scale requirements for efficient production. The most outstanding example of this trend has been the emergence of the mini-mill, a source of new competition that cannot be blamed on the vexed issue of international trade.

In general, the American steel industry has not responded to its drastically altered situation by radically revising its behavior or its strategies. Although inadvertent, the industry's most effective response has been to alter its structure, shifting capital toward mini-mill firms. The strategic ineptitude of the traditionally dominant integrated sector stems from a basic misunderstanding of the secular forces at work, emphasis on oligopolistic rivalry for market share, and ingrained resistance to change. These weaknesses have been compounded by an increasingly strong belief that governmental policies are the ultimate determinants of industry performance. While this may be an arguable position in a very broad sense, the industry has acted upon this view by applying a far more dubious corollary: that the effort to alter policy offers more potential benefits than risking fundamental changes in the industry's strategy, behavior, and performance. In the United States, both industry and government have generally resisted change, since both parties have drawn some partial benefits from the maintenance of the status quo. This contrasts sharply with the progressive interaction between government policy and industry strategy in several other countries.

The dynamism of the mini-mill sector provides encouraging evidence that there is still a strong progressive component in the U.S. steel industry and that this industry can be highly competitive internationally. In order to revitalize the integrated sector of the industry, however, radical surgery will be required. The forces of decline have been at work for several decades; this alone suggests that they will not be reversed by marginal or gradual changes. A similar conclusion applies to several other basic industries in the United States, and this is one reason that the steel industry provides such an important case study. If there are few basic industries in which decline has been so protracted, there are fewer still in which the emergence of a dynamic sector, based on different technology and a different managerial philosophy, so clearly indicates the lineaments of a restructured and healthy industry. In this concluding chapter, we will suggest how this inevitable transition can be eased by changes in the industry's strategy, the industry's structure, and the government's policy.

THE STRATEGY-POLICY NEXUS: LESSONS FROM THE PAST

On the most fundamental level, an industry's performance is determined by the economic conditions which characterize its market: the presence or absence of growth in demand, the availability of funds for investment, the price of inputs, the existing physical plant, and so forth. While it would be foolish to dismiss such factors, it is also the case that their implications can be radically altered by appropriate governmental policies and managerial strategies. Objective weaknesses can be transformed into strengths by the right mix of strategy and policy, while the wrong mix can dissipate seemingly unassailable advantages. The potential for a complementary and progressive mix of strategy and policy—as well as the difficult and fragile character of such a balance—can be illustrated by summarizing the Japanese, Canadian, and U.S. experiences during the postwar period.

Japan

In the first years after World War II, the prospects for Japanese steel producers seemed extremely bleak. Existing plants were small and inefficient, so that the industry was only marginally competitive in its home market. Labor costs were low, but other inputs were expensive; Japan lacked both the required raw materials and the capital needed for investment. Nevertheless, both the Japanese government and the business establishment regarded a strong domestic steel industry as an essential part of a strong postwar industrial economy, and this shared perspective produced a cooperative meshing of government policy and industry strategy. This consensus progressively solved the problems which were inherent in the initial postwar condition of the Japanese steel industry.

From the first years after the war, the industry recognized that its success would depend on its ability to reduce costs, given world market prices. Cost reduction required both increased efficiency and lower input prices. On the efficiency front, the Japanese industry sought to emulate its far more efficient American counterpart by substituting capital for labor, although it was handicapped by limited capital availability. To reduce this constraint, the government initially provided most of the investment funds, established generous tax benefits, and encouraged specialization. Particular attention was paid to improving efficiency in the use of imported iron ore and coal and to boosting labor productivity. Higher efficiency reduced the impact of high materials prices, a problem which was also addressed via temporary government subsidization of imported supplies, long-term contracts with low cost producers, and efforts to reduce transportation costs. Eventually, this strategy proved a dynamic success: lower production costs solidified the industry's position in its home market and set the stage for rapid entry into export markets.

By the mid-1950s the Japanese steel industry had eliminated its original cost disadvantage; its success encouraged it to refine its strategic approach. With a rapidly growing home market and increasing foreign sales, Japanese steel producers pursued operating efficiencies and lower unit capital costs via the greenfield route, constructing enormous state-of-the-art plants at virgin sites. These plants were uniformly located on the coast in order to facilitate the importation of raw materials and the export of finished products. They were built ahead of the market, on the assumption that rapid market growth would eliminate surplus capacity and lead to continual expansion of the scale frontier. Raw materials problems were reduced by developing new sources of supply in countries like Brazil and Australia and by building large bulk carriers to lower transportation costs. Labor costs were controlled by rapidly improving productivity, by offering lifetime employment in exchange for moderate wages, and by contracting out many activities.

Once the problems of raw material availability, capital scarcity, and excessive costs had been resolved, the direct governmental role receded, although the Ministry of International Trade and Industry (MITI) continued to provide guidance and counsel. When demand lagged, MITI encouraged the establishment of cartels to minimize aggregate losses; when investment plans diverged from the industry-government consensus, MITI attempted to preserve collective discipline. In general, though, the industry's greenfield strategy was self-sustaining, bringing stupendous improvements in performance and lower costs, which continually expanded the markets in which the Japanese were competitive.

By the early 1970s, the felicitous combination of industry strategy and government policy had produced the most efficient steel industry in the world, with technically advanced and optimally sized facilities. Japan's best plants were third-generation greenfield giants. Reliance on export markets, which had initially been regarded as a source of foreign exchange, became a means of maintaining high operating rates and profitability. The cost competitiveness of the Japanese industry gave it some measure of pricing power, but this potential was used to penetrate foreign markets and to provide low-cost steel to Japanese industries rather than to raise prices.

In spite of such success, the post-1974 steel crisis and the emergence of global excess capacity brought changes that have undermined the strategy that had served Japanese steel producers so well. Home market growth has faltered, while hard-pressed foreign producers have clamored for protection against the more efficient Japanese. As a result, political factors have kept the Japanese from exploiting their cost advantage. The initial response of the Japanese steel industry to these changed conditions was to shift its orientation from "efficiency through expansion" to the rationalization and specialization of existing capacity. Lagging exports were supplemented by sales of technology, including the turnkey construction of entire plants in developing countries. Such sales were made despite MITI's reluctance—a sign that the smooth consensus between government and industry was also disturbed by changing conditions.

While the Japanese industry has had greater success than its competitors in coping with the changed conditions of the 1970s and 1980s, the long-term prospects of the industry have dimmed considerably. The strategic orientation that characterized the heroic period of the Japanese industry seems ill suited to the market conditions of the 1980s. The persistence of worldwide excess capacity is an extremely ominous development for an industry in which operating rates depend on volatile and highly politicized export markets. Lower operating rates generate lower profits, straining the highly leveraged balance sheets of Japanese steel firms and exposing the Achilles heel of massive facilities. In weak markets, scale dependent efficiencies vanish, since low operating rates increase unit capital costs and decrease labor productivity. The changed market conditions of the 1970s and 1980s have demonstrated the risks of the Japanese greenfield strategy. Regardless of their hypothetical efficiency, plants designed according to the old strategy are incompatible with the greater flexibility required under contemporary conditions. A new strategic orientation is needed.

Canada

In spite of vastly dissimilar backgrounds, both the Canadian and Japanese steel industries suffered from similar weaknesses in the early 1950s: a small home market (although with some growth potential), high costs, outmoded facilities, and limited capital availability. If anything, Canadian circumstances were even less promising than those in Japan. For both demographic and geographic reasons, the underlying potential of the Canadian market was inferior, and import penetration was substantial under normal circumstances.

In the Canadian case, these conditions elicited a strategy radically different from the one developed in Japan. Its underlying principle, that the Canadian industry would be a price taker, was the same, as was the corollary that the industry would therefore have to concentrate on lowering costs. But whereas the eventual Japanese response to this condition was to seek the maximum economies of scale by building ahead of the market, the Canadian approach was to specialize and to lag behind the market in order to maintain the highest possible operating rates. High operating rates would lower costs and increase profits, generating funds for investment. Thus, given a limited home market, high operating rates rather than major economies of scale were viewed as the most appropriate means of offsetting high capital costs and therefore initiating a virtuous cycle of cost reductions, higher profits, and greater investment.

The Canadian concentration on cost competitiveness implied developing only those markets where the Canadian steel industry could be a low-cost supplier. Hence, certain geographical markets were surrendered to imports, while exports were sought only in "natural markets"—e.g., across the Great Lakes. Product lines in which economies of scale were less significant were left to mini-mills, while the integrated producers maximized scale in a limited market by specializing to reduce capacity redundancies.

As was the case in Japan, the implementation of this strategy involved the emergence of a consensus between government and industry (although the consensus was less explicit in Canada). Most crucially, the government accepted that profitability was the key to success, a view that had three major corollaries in terms of policy. First, tax rates were altered to improve industry cash flow, although such changes were not steel specific. Second, the liability of a small domestic market and the importance of specialization were recognized. Third, the Canadian consensus included the use of foreign trade as a buffer to sustain high domestic operating rates. When supplies were tight, the import spigot was in effect opened wider; when demand faltered, imports were in effect pushed out. For their part, Canadian steel firms committed themselves to supplying the domestic market with low-priced steel even under conditions when export markets might be more attractive.

The Canadian strategy proved highly successful; by the mid-1970s, the Canadian steel industry was the most profitable in the world. Furthermore, the strategy pursued by the Canadian industry appeared well suited to the market conditions of the late 1970s and 1980s, when slow market growth made high operating rates essential. Perversely, in the late 1970s the Canadian steel industry broke with the strategy that had served it so well from 1955 to 1975. The industry's relationship with the government turned sour; capacity was added ahead of the market, forcing Canadian steel into the export market. Thus, as was also the case in Japan, the Canadian industry entered the 1980s facing a much more uncertain future than it had foreseen just a few years before.

The United States

In the case of the U.S. steel industry, one searches in vain for the kind of coherent, realistic, and constructive strategy that was so characteristic of the Japanese and Canadian steel industries through most of the postwar period. One reason for this is that the relative prosperity of the U.S. industry in the late 1940s and 1950s did not produce an environment in which both the industry and the government agreed that significant changes were required. Whereas the Japanese and Canadians were able to construct effective strategies around the goal of fostering improvement, the focus in the United States has been on resisting decline—a very different goal. The U.S. industry was unable to develop a strategy for retrenchment—a process inherently more difficult to manage than expansion. Instead, it has sought to maintain a disappearing status quo.

From 1901 until at least 1960, the American steel industry was dominated by a relatively stable oligopoly, the key feature of which was price maker behavior.[2] In terms of management strategy and conduct, the price maker perspective of the U.S. industry is the feature that most clearly distinguishes it from its Japanese and Canadian counterparts. Prices in the U.S. steel industry were usually based upon some markup over costs; price changes were infrequent and generally followed clear shifts in costs. The U.S. Steel Corporation, as the largest firm, defined the limits of the industry's structure and managerial culture. Other firms devised their strategies under the U.S. Steel umbrella; so long as their costs were below those of Big Steel, they could prosper. In this environment, price competition was eschewed as disruptive, and the industry's leader ensured that "price chiseling" was discouraged. U.S. Steel allegedly never discounted its list prices, regardless of the level of demand, until the late 1960s.[3]

Given the industry's oligopolistic structure, rivalry among integrated steel producers deemphasized price competition in favor of jockeying over market share. This implied that the successful firms were those which served growing markets; all firms, however, were committed to protecting (or expanding) market share during boom periods. Since these were infrequent, low operating rates prevailed over most of the business cycle. The tendency to excess capacity was enhanced by a generally overoptimistic view of potential market growth, based on a short-term perspective that emphasized the rare shortages and that blamed poor performance on trade problems. The more fundamental secular forces—such as declining steel intensity, evolving product markets, and changes in comparative costs—were as a rule ignored.

The industry's price maker behavior emphasized setting prices, given costs, rather than controlling costs, given prices. American integrated steelmakers did make efforts to lower costs, but these efforts were based on an inappropriate strategy. This strategy, which we have termed massive modernization, involved the replacement of existing capacity through the construction of ultralarge facilities, attaining full economies of scale and producing a more or less complete line of products. Due to their reliance on this strategy, seemingly confirmed by Japanese success, integrated firms have been unprepared to cope with the emergence of mini-mills, which embody a very different strategic and technological approach. Furthermore, inadequate rates of return on investment and slow market growth have barred U.S. integrated firms from all but a partial implementation of the approach to which they have committed themselves.

The problem with this approach is not that a lack of funds thwarts the massive-modernization strategy but rather that the inappropriateness of this strategy ensures that funds will not be available to implement it. The U.S. industry has sought to run with the Japanese, but without the market growth that has been the driving force of superior Japanese performance. U.S. integrated producers have several large new facilities that are the equal of Japanese equipment—for example, the new blast furnaces and coke ovens at Inland's mill and at Bethlehem's Sparrows Point plant. Unfortunately, in most cases such new facilities are modern islands in an increasingly aging sea. Moreover, the enormous capital costs of such facilities largely offset the performance benefits they provide. Thus the effort to implement the massive-modernization strategy has exacerbated the divergent trends in the performance of U.S. integrated producers and their major competitors, both foreign and domestic.

The industry's response to the paralysis induced by the massive-modernization strategy has not been to question the underlying logic of this approach. Instead, steel firms have sought major revisions in government policy in order to boost the funds available to them—whether through trade barriers (which increase market share and raise prices), changes in tax policies, reduced regulatory expenditures, or other measures. The politicization of the industry's efforts to restore its competitive position reflects the fact that steel prices have always been a political issue; recent efforts to improve cash flow via political actions have often received a sympathetic hearing in Washington. For example, in the waning years of the Carter administration, the Steel Tripartite Committee—a body comprising representatives from industry, labor, and government—determined that the American steel industry needed to invest more than $5 billion annually (in 1980 dollars) to maintain competitiveness. While this was somewhat less than the industry's own estimates, it reflected the shared assumption that the massive-modernization route is the only path to a revitalized American steel industry.[4]

While integrated firms, the union, and political figures have idealized this strategy, financial markets have recognized its real effects. Due to high capital costs, the substitution of capital for labor has failed to reduce real expenditures, even though high labor and energy costs in the 1970s made this even more urgent. Mini-mill technology has stepped into the breach, providing significant savings in labor and energy at modest capital costs and reducing total expenditures compared to integrated techniques. Mini-mills have pioneered ways to economize in construction by building simply, quickly, and cheaply, and they have now usurped the fast track in steelmaking technology. Integrated producers have learned neither how to adopt capital-saving technologies nor how to build economically using current technology. As a result, the cost gap between the mini-mill and integrated sectors has been widening. The profitability of the mini-mill sector has made it relatively easy for such firms to attract capital, while the financial performance of the integrated sector has been burdened by doomed attempts to implement massive modernization. Cost competitiveness is becoming an ever more crucial requirement in the steel industry just as the cost reduction strategy preferred by integrated firms is collapsing.

The integrated sector's passive attitude toward costs has a rather surprising corollary, namely, an indifference toward profitability. Despite the fact that poor profitability has long been the industry's major complaint *in the political arena*, its record is fraught with examples of how its own actions have undermined its rate of return. Excess capacity has been maintained in order to protect market share for the rare booms, reflecting a "tons mentality" rather than profit maximization. Integrated firms have sought to remain veritable supermarkets of steel products, with little regard for the actual costs and rates of return in specific product lines. Given these commitments, capital expenditures have been widely dispersed, reducing the efficiency of investment. Finally, relatively high input prices have been tolerated due to excessive vertical integration (encouraging the use of high-priced domestic ores) and labor contracts that have sought to enlist the union as a political ally. Failures in these areas reflect the complex links among the industry's oligopolistic traditions, its price maker behavior, its failure to minimize costs, and its political focus.

This complex and counterproductive pattern is now crumbling under the blows of increased competition, both foreign and domestic. Most crucially, the traditional steel oligopoly has disappeared, a development that can be identified with the decline and fall of the U.S. Steel Corporation. The oligopolistic status quo was viable only so long as the industry leader retained its standing, and that has now been lost. U.S. Steel's market share has fallen to the point where it is at best first among equals. Its purchase of the Marathon Oil Company represents its abdication from the throne—if not its exit from the steel industry. While the implications of this event have been masked by the steel crisis of 1982–83, they are profound. Most significantly, perhaps, it indicates the urgency of defining a new strategy for the American steel industry. For several decades, the industry's principal strategic goals have been defensive. Integrated firms have sought to protect the traditional structure of their industry and their price maker status, but there is now precious little to protect.

ECONOMIC CONDITIONS IN THE 1980S AND 1990S

The first step toward the development of a more effective strategy is a sober assessment of the forces affecting the future development of the steel market. Firms must foresee looming changes in the industry and position themselves accordingly. U.S. producers should now be making decisions based upon their assessment of where the market will be in the year 2000—what the overall level of demand will be, what product markets will be strong, and who their major competitors will be.

Steel Demand

Regardless of the particular problems of the steel industry, the basic conditions faced by that industry through the rest of this century will inevitably reflect the broader forces reshaping the U.S. economy. As was described in Chapters 3 and 4, the structural changes in the U.S. economy all suggest a continued and accelerated decline in steel intensity. Steel intensity has been declining during most of the postwar period, and this trend has accelerated since the energy crisis. As a result, high rates of real GNP growth (approximately 2% per year and above) will be required to forestall actual declines in steel consumption.

The declining secular trend in steel intensity has been exacerbated by the structural changes that are currently reshaping the U.S. economy. Many of these changes were described in Chapter Four; on the most general level, they are connected to the technological revolution in advanced electronics, the internationalization of the U.S. economy, the changes in relative prices associated with the energy crisis, and an upward shift in real interest rates. Such structural changes portend an even more rapid decline in the overall economic significance of steel production; some reasons for this are listed below:

- Steel consumption in the major steel market, the automobile industry, will be well below the levels of the late 1970s. One reason for this will be continued efforts to reduce automobile weight. Lower sales of domestically produced vehicles relative to GNP growth may be even more significant.
- As in the automobile market, indirect steel imports (i.e., imports of goods made from steel) will grow in importance, reducing the domestic demand for steel.
- The substitution of aluminum and other lighter materials for steel (e.g., in containers) is a well-established trend that can be expected to continue.
- Manufacturing will shift away from methods utilizing heavy machinery to less steel-intensive electronic methods, reflecting significant decreases in the relative price of electronic equipment.
- An ever greater share of capital spending will be devoted to economic activities that require less steel per unit of value than more traditional investment outlets. For example, from 1976 to 1981, real investment in office equipment increased by almost 300 percent. More traditional steel-intensive investments (fabricated metals, metalworking machinery, general industrial equipment, etc.) grew hardly at all in real terms.[5]

- Consumer tastes will shift away from steel-intensive appliances (refrigerators, freezers, etc.) toward electronic appliances (computers, video equipment, etc.), reflecting changes in relative prices and the availability of new products as a result of technological progress.

Figure 10–1. Projected Steel Consumption by Region (millions of net tons, crude steel equivalent).

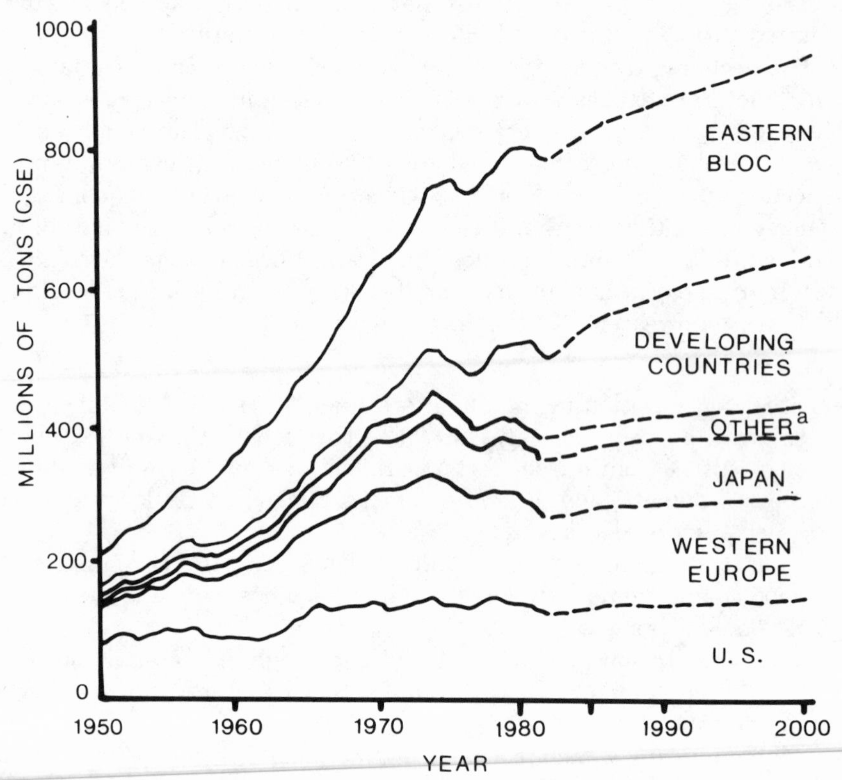

a. Other developed market economies (Canada, South Africa, Australia, New Zealand).
Sources: World Steel Dynamics, *Core Report R* (New York: Paine Webber Mitchell Hutchins, 1981); Organization for Economic Cooperation and Development (OECD), *The Steel Market in 1981 and the Outlook for 1982* (Paris: OECD, 1982), International Iron and Steel Institute (IISI), *Steel Statistical Yearbook* (Brussels: IISI, various years); and authors' estimates.

Thus, the first and foremost element in the steel market of the 1980s and 1990s will be the persistence of slow growth in steel consumption. Figure 10–1 describes our projections of world steel consumption, disaggregated by region, through the rest of this century. Because of the persistence of declining steel intensity and the economic dislocations associated with the structural transformation of the U.S. economy (which suggest relatively poor GNP growth on a trend-line basis for several years), we project no significant increase in U.S. steel consumption through the rest of this century. Due to the existence of similar conditions in other developed countries, the same pattern is evident there as well. While this may seem excessively conservative, Figure 10–1 shows clearly that this forecast is not at variance with the overall postwar trend in U.S. steel consumption.

The general weakness in steel demand will be more prevalent in some markets than in others, so that structural changes will persist. The demand for mini-mill products (bars, rods, and light structural elements) will be relatively strong, while flat-rolled products will be most affected by declining steel intensity. Certain flat-rolled categories (e.g., coated products) will gain new markets, but this will not offset the overall decline in this product line.

Finally, consumer needs will encourage the further fragmentation of the steel market, a process that is already fairly advanced. Increased diversity in their product requirements will force consuming industries to be more demanding in terms of steel quality (weldability, ductility, metallurgical properties, consistency, etc.). Mini-mill operations will expand the markets in which they are competitive, forcing integrated producers to specialize in more sophisticated products of high quality and high value. The ongoing fragmentation in steel markets will also boost the role of the steel service centers. Besides holding larger inventories of an ever wider variety of products, the service centers will extend their productive activities into more finishing operations.

Steel Supply

The 1960s introduced two new players in the U.S. steel market: foreign producers and mini-mills. Future developments will almost certainly enhance the role of the mini-mill sector. While the outlook for imports is clouded by political considerations, there is little doubt that imports will continue to challenge domestic producers with intense and effective competition.

To some extent, the prospects for imports into the U.S. steel market vary inversely with the state of world steel demand. Steel consumption in developed countries will grow slowly, if at all, during the 1980s and 1990s. In the developing world, where steel intensity is still rising, consumption could grow at more than 3 percent per year. The projections shown in Figure 10–1 suggest that excess capacity, a burden already carried by most developed countries, will persist for some time. Retirement of facilities could redress this imbalance, but there is little evidence of this as yet—Britain being the most obvious exception. "You first" seems to be the prevailing attitude of steel firms and steel-producing countries toward the elimination of excess capacity; this stance only perpetuates the abysmal state of the market. As capacity reductions occur in developed countries, these will be at least partially offset by the maturation of steel industries in the Third World. Thus the internationalization of the steel industry will continue. Widespread trade in steelmaking technology will tend to standardize efficiency levels, while excess capacity and the elimination of noncompetitive sources will tend to set one world price for raw materials (although significant differences in transportation costs will perpetuate some price differentials). Labor costs, especially relative to the national manufacturing average, will play an increasing role in determining competitiveness—a fact suggesting that U.S. firms may face even more difficult times.

In the absence of trade restraints, these considerations argue that import penetration of the U.S. market will continue, at the expense of domestic integrated producers. In particular, by the year 2000 Third World suppliers will be highly competitive in commodity products such as standard plates, structural shapes, and hot- and cold-rolled sheets. This is due in part to the fact that developing countries are the only ones that will enjoy the market growth that has in the past determined trends in the global distribution of steel production and the shifting pattern of international competitiveness. Increased scrap prices are unlikely to reverse the cost advantage of mini-mills, so that they should be able to maintain or expand market share in their products vis-a-vis imports. This is not likely to be the case for integrated producers.

Financial Constraints

Besides the changes that will directly affect their market, steel firms will also be forced to cope with significant shifts in the macroeconomy, especially in financial markets. Until 1980, extremely low (and sometimes negative) real interest rates were a consistent feature of the postwar U.S. economy, and they have in effect subsidized unprofitable firms and industries like integrated steel. These low real interest rates have been part and parcel of an overall inflationary environment. Low real interest rates, reflecting individuals' shortened time horizons, have reduced savings and disrupted the valuation of investments. Inflationary increases in the money supply have artificially maintained the pool of funds available for investment, but the prosperity this provides has become increasingly fragile.

High real interest rates will probably be needed to boost the savings rate in the United States; thus, firms are likely to face a higher real cost of capital for the foreseeable future. At the same time, the demand for funds will also increase due to the advanced age of many industrial facilities, the pressure of international competition, and the availability of significant new technologies. These demand side factors (not to mention the potential for significant government borrowing) will also tend to bid up interest rates.

As a result, the true opportunity cost of capital will play a much larger role in rationing funds among competing uses. This is an ominous development for American steel firms, since their return on investment has been so poor. Poor rates of return on investment, which are barely acceptable when real interest rates are low, will be unsustainable when interest rates reflect the true value of capital resources. As the criteria for profitable investments become more stringent, many traditional steel projects—at least those favored by integrated firms—will be denied funding. Although the industry has complained of a capital shortage for years, it has actually spent huge sums of money on relatively unprofitable investments: Republic's commitment to coke ovens instead of continuous casters, U.S. Steel's mistaken efforts to compete in wire rod markets, and so on. This is a pattern that financial markets are unlikely to support in the future; as a result, capital constraints will tighten.

SURVIVAL STRATEGIES: THE STRUCTURE OF A COMPETITIVE INDUSTRY

Weak markets, increased competition, and severe capital constraints describe the three key features of the environment that will confront the U.S. steel industry during the rest of this century. In these circumstances, integrated producers will not possess significant market power; indeed, their traditional price maker status has already broken down, except insofar as it has been propped up by trade barriers. It is also clear that continued adherence to a massive-modernization strategy is a recipe for disaster. Future profitability depends on being attuned to customer requirements and on seeking to be the low-cost supplier to the markets in which firms choose to compete. In the 1980s and 1990s, being big will be less significant than being quick—quick to adapt to the market, quick to exploit new technologies, and quick to pare unprofitable operations.

Faced with such challenges, integrated firms will increasingly seek to diversify away from steel production. While this may represent the optimal strategy for integrated steel *firms*, at least in terms of protecting the value of shareholders' assets, it does not a represent a strategy for the steel *industry*. We will not discuss diversification; instead, we will describe the key elements in a viable strategy for firms that remain committed to the steel market.

A New Strategy for the Integrated Sector

Under these conditions, a strategy like the one adopted by the Canadian steel industry from the mid-1950s to the late 1970s offers a far better model for U.S. producers than does the greenfield strategy of the Japanese. The most relevant feature of the Canadian experience is the commitment to high operating rates. These are crucial to the revitalization of the American steel industry, given the poor prospects for significant market growth. In the short run, operating rates depend upon the level of demand, and this encourages the perception that firms have no control over this variable. In a dynamic sense, however, rates of capacity utilization are determined by managerial decisions; this is clearly suggested by comparing historical operating rates in the U.S. and Canadian steel industries (Table 8–2).

Retrenchment and Specialization. Low operating rates are a characteristic feature of the U.S. steel industry—a legacy of the period when price competition was weak and when excess capacity ensured the maintenance of market share during boom times. As a result, U.S. producers have tended to target capacity levels toward the boom, even though this weakens profitability and cost competitiveness at other times. The competitive conditions under which excess capacity was acceptable are now gone; firms must seek to run at an attractive operating rate (ca. 90%) in times of normal demand rather than only during the boom. While this may mean some loss of potential tonnage during periods when demand is strong, increased profitability through the entire cycle would more than compensate for this. Individual firms may also improve operating rates by striving to become low-cost suppliers to the market and aggressively cutting margins when demand lags—displacing imports and higher cost domestic producers. This approach has frequently enabled mini-mills to avoid severely depressed profits during recessionary periods (at the expense of integrated producers and imports).

Excess capacity represents a waste of capital; and capital savings, rather than reduced operating costs, may be the more significant benefit provided by high operating rates. Running tight—that is, forgoing the cushion of slack capacity maintained for the boom—reduces expenditures on redundant equipment, disperses maintenance funds over a narrower range of plants, and raises the efficiency of investment through the synergistic benefits of concentrating capital expenditures at fewer sites. Since capital constraints will be increasingly severe, firms that fail to counter this trend by raising operating rates (therefore reducing capital costs) will find it difficult to compete.

Specialization by product is an essential complement to capacity reductions at the plant level. Specialization permits a firm to achieve significant economies of scale without investing in the large, complex plants that characterize the massive-modernization strategy. Mini-mills and foreign producers (increasingly from developing countries) will be the low-cost suppliers for basic commodity products. This makes it imperative that integrated firms realistically define the markets in which their costs will be lowest (or, alternatively, successfully adopt mini-mill techniques for mini-mill products). Most flat-rolled markets will continue to be supplied by integrated producers due to the significant economies of scale in the hot strip mill; the same is true (albeit to a lesser extent) for heavy structurals, heavy plates, and similar products. Rather than attempt to resist mini-mill incursions in other product lines, integrated firms would do better to concentrate on products where economies of scale and other factors favor traditional integrated techniques.

Specialization by product could also be linked to specialization by process. Many American steel plants have a mixture of good and bad facilities—a legacy of the dispersed pattern of investment. The cost of thorough modernization, even with capacity reductions and a narrower product range, would still be extraordinarily high. As an alternative, the dedication of plants to a few processes (while other facilities are closed) and the exchange of intermediate products—such as coke, hot metal (over short distances), semifinished shapes, hot-rolled coil, and cold-rolled coil for galvanizing—could encourage the operation of only the best facilities. Some American locations, such as the Chicago area (six integrated plants and one major finishing facility within fifty miles of one another) offer almost ideal prospects for rationalization by process.

Specialization by process provides one final advantage when combined with specialization by product. Targeting a limited product range involves the risk of excess capacity in market downturns; at present, this risk is reduced by the ability to shift among product lines. Downside risks are usually borne most heavily by those facilities where economies of scale are greatest (e.g., blast furnaces and hot strip mills), while the costs are limited for facilities where economies of scale are less significant (e.g., most finishing facilities). Through specialization by process as well as by product, the problem of excess capacity in market downturns could be reduced for facilities where economies of scale are important. Such facilities could supply a variety of mills, each serving different markets.

Specialization by process should be perceived as a stopgap measure rather than as a long-term solution for integrated steel producers. Process specialization does offer the advantage of concentrating capital expenditures while overall integrated capacity is being reduced, although it does so at the cost of higher operating costs due to the increased handling required. In the long run, however, the key strategic imperative is the reduction of total capacity and the transition to leaner, more flexible plant oriented toward being the low-cost suppliers to selected markets.

In the final analysis, the market will determine the future structure of integrated steel plants. The successful integrated producers will be those who anticipate the direction in which their market is heading, so that they will be able to grow within a basically stagnant market (as the mini-mills have done). In our view, integrated firms must concentrate on higher value and higher quality products tailored to increasingly diverse customer requirements (in metallurgy, coatings, size, etc.)—perhaps involving whole new generations of steel products (e.g., plastic bondings, graphite laminates, etc.). Such a high-value, high-quality orientation seems suited to America's comparative advantages, although it conflicts with the tonnage mentality that has prevailed within the steel industry. The increasing diversity of customer requirements may shift finishing processes (coating, tempering, etc.) toward the market, that is, to the steel service centers.

Thus the market is reducing integrated steelmaking to its core strengths, the operations where economies of scale are significant: ironmaking, steelmaking, and the hot rolling of flat-rolled products. Even these operations, however, will have to be increasingly specialized to meet divergent finished product requirements. At the same time, integrated plants will have to become more flexible, quickly adopting continuous technological refinements in both processes and products. The leaner, cleaner, and hopefully meaner plants that emerge will comprise facilities that attain the minimum optimum scale but that are attuned to high operating rates rather than to massive capacity (e.g., blast furnaces that produce 5,000 rather than 10,000 tons per day). This approach will certainly mean less ownership of iron and coal properties. Vertical integration is likely to be less attractive than horizontal integration, possibly with competing technologies (such as plastics and aluminum) or with foreign producers such as the Japanese. Such realignments are already occurring: U.S. Steel, for instance, is seeking foreign participation at some of its plants, and Japanese steel producers have evinced some interest in purchasing U.S. mills.

Dynamic Cost Reduction. While the Canadian industry offers some valuable lessons to integrated producers in the United States, an even richer model is provided by the mini-mills, which are in effect the market's response to the postwar conditions faced by American steel producers. Product specialization, aggressive price competition, high operating rates, and a market orientation are all characteristics of mini-mill operations. Just as significantly, the mini-mill phenomenon reflects not just a more successful marketing strategy but also the progressive effects of technical change. The commitment of mini-mills to the rapid adoption of appropriate technologies has enabled them to sustain the dynamic cost reductions that are essential to the revitalization of the integrated sector.

The transition to price taker status makes it essential that steel firms be able to sustain dynamic cost reductions, not the halting and ambiguous benefits associated with the massive-modernization strategy. The mini-mill sector is a suitable model for integrated firms, particularly in regard to the importance of capital-saving technology and operating techniques–i.e., operating practices and incremental technological changes that are not contingent on enormous capital expenditures. Since the key to cost reduction is advanced technology rather than massive scale, firms should define their investment goals accordingly.

Outlays for expensive facilities that do not embody significant technological improvements (e.g., coke ovens) are in a sense wasted, especially in a period of capital constraints. It would be much better to contract for deliveries of materials like coke and to devote capital expenditures to processes where real technological breakthroughs are possible. The most attractive technologies are those that not only lower labor requirements but reduce capital needs as well; the simple conception of substituting capital for labor–the underlying assumption of the massive-modernization strategy–is a dangerous one. Continuous casting (replacing ingot casting and primary rolling) and the electric furnace (replacing the entire complex of coke ovens, blast furnaces, etc.) are the best examples of capital-saving steel technologies; together, they constitute the core of the modern mini-mill. In the future, similar gains may be obtained from advances in direct rolling (eliminating reheating furnaces and soaking pits), direct reduction of iron ore, plasma steelmaking, and so on.[6]

Capital constraints imply that ways must be found to improve efficiency without major expenditures, and this focuses attention on improved operating techniques, which have greater cumulative effects than major innovations. Improved operating techniques are essential in at least five areas. The first is the more thorough collection and monitoring of operating statistics. The basic productivity data discussed in Chapter 5 are readily available to Japanese firms, so that they are able to identify the processes in which their own performance is weak and to alter their practices or investments accordingly. U.S. firms, by contrast, often have only vague perceptions of their actual performance—a situation that naturally makes it more difficult to sustain a rapid rate of improvement.

Second, integrated firms should emulate their mini-mill competitors in terms of their attitude towards new construction. Engineering staffs at integrated firms must break with their traditional preference for monumental, overly complicated, and expensive facilities. Facilities should be designed to be replaced as they become outmoded rather than to last forever. This also implies that integrated firms should seriously investigate the structure of scale economies for those products in which they specialize and should seek to attain the *minimum* optimal scale, an approach that makes it easier to maintain high operating rates.

A third requirement for improved operating techniques is a more constructive relationship with the labor force. While unit labor costs are currently a severe competitive disadvantage for U.S. producers, reductions in hourly compensation like those achieved in the March 1983 contract do little to alleviate this. Negotiated savings in hourly employment costs provide steel firms with a static benefit; there is no guarantee that these benefits will not be lost in the next contract, should demand then be strong. A more long-term approach, concentrating on improving productivity through altered work rules and increased reliance on the insights of production workers, would offer dynamic benefits without relying on capital expenditures to generate productivity improvements.[7] Unfortunately, labor relations in the American steel industry have tended to be highly authoritarian—a pattern that will be hard to break. Nevertheless, this is at least as important as reducing the differential between steel employment costs and the manufacturing average.

Fourth, improved operating techniques entail constant technical refinements in equipment. When tied to the goals of specialization in high-value products, this suggests that integrated firms must devote much greater attention to the finishing end. Integrated producers should modernize backward, from the market, rather than the traditional orientation of forward, from the raw materials. Concentration on the finishing end also implies that quality-oriented investments should get a large share of limited capital expenditures. In particular, greater use of computer controls is essential, since computerization should be a comparative advantage for U.S. producers relative to many of their foreign competitors.

Finally, if technology is the key to dynamic cost reductions, integrated firms must boost their expenditures on research and development and shift such activities more toward process research (although this should not entail reductions in product research). American steel firms have recently begun to adopt foreign technologies, and this is a positive step, especially insofar as technologies purchased off the shelf are less expensive than those designed in-house. Yet total reliance on purchased technologies would be a particularly dangerous state of affairs for U.S. producers given their labor-cost disadvantage. This implies that the adoption of foreign technologies must be supplemented by internally developed innovations. Given the present constraints faced by U.S. steel producers, potentially capital-saving technologies should be the focus of most process research.

In fact, research in new process technologies has the potential of permitting a streamlined U.S. industry to skip a generation in steelmaking technology. Paradoxically, the industry's lack of investment in recent years may prove to be an advantage in the long run—provided profitability can be restored by the strategic orientation outlined above. While American steel producers and even outside observers have been urging an effort to match the Japanese, technological and market trends have been undermining the viability of the Japanese approach. In a sense, the American industry is fortunate that it has no new greenfield plants. In Europe, government funds were often used in the late 1960s and early 1970s to mimic the Japanese strategy. The result has been substantial excess capacity, massive losses, and further government intervention. Such plants are the product of a 1950s mentality; different standards will apply in the 1980s and 1990s.

Under present market conditions, the massive-modernization strategy is fraught with contradictions. Large scale provides hypothetical economies that market demand will not support; significant improvements in physical efficiency are dissipated by extremely high construction costs; monumentalism freezes technology and product mix; and so on. In contrast, the strategy that we have outlined here is coherent: its goals reinforce each other and correspond to the underlying trends in the U.S. steel market. The core of this strategy is to eliminate excess capacity in order to raise operating rates; this will lower operating costs, increase profitability, and reduce capital requirements. Capacity reduction can and should be tied to product specialization—especially in more sophisticated products, where growth prospects will be strongest. Specialization lowers the minimum efficient scale, indirectly reducing capital requirements.

Table 10–1. Trend Projections for the 1980s and 1990s.

	1980[a]	*1985*	*1990*	*2000*
Apparent steel consumption (mm t)	101.1	103.0	107.0	117.0
Import share of U.S. market (%)	17.9	22.5	25.0	25.0
Domestic shipments (mm t)	83.0	79.8	80.2	87.8
Imports (mm t)	18.1	23.2	26.8	29.3
Exports (mm t)	2.8	2.5	2.5	2.5
Total shipments (mm t)	85.8	82.3	82.7	90.3
Mini-mill share of total shipments (%)	13.8	19.2	25.1	34.6
Mini-mill shipments (mm t)	11.8	15.8	20.7	31.2
Mini-mill yield	0.84	0.86	0.89	0.93
Mini-mill raw steel production (mm t)	14.0	18.4	23.3	33.6
Mini-mill capacity (mm t)	16.0	20.5	25.9	37.3
Mini-mill productivity (MHPT)	4.0	3.3	2.6	1.8
Integrated[b] shipments (mm t)	74.0	66.5	62.0	59.1
Integrated yield	0.73	0.75	0.79	0.87
Integrated raw steel production (mm t)	101.8	88.7	78.5	67.9
Integrated capacity (mm t)	138.4	105.5	91.1	75.5
Integrated productivity (MHPT)	9.5	8.3	7.0	5.0
Total raw steel production	115.8	107.1	101.8	101.5
Capacity Utilization Rate (%)	75.0	85.0	87.0	90.0
Total capacity (mm t)	154.4	126.0	117.0	112.8
Total productivity (MHPT)	8.8	7.3	5.9	3.9
Total employment[c] (000)	396.3	316.2	256.8	185.4

a. Based on five-year average around 1980.
b. "Integrated" includes specialty steel production throughout this table.
c. At 1,900 hours worked per year.

Retrenchment and specialization, however, are only the first steps; they must also be linked to a greater commitment to dynamic cost reduction. This involves thorough monitoring of performance, greater modesty in design, more constructive labor relations, rapid adoption of breakthrough technologies (especially those of the capital-saving type), and greater commitment to R&D. All of these measures are designed to establish an environment that generates dynamic cost reductions. Such a result is the key to restoring the industry's profitability—and profitability, rather than tonnage or market share, will determine the future of American steel firms.

Restructuring the American Steel Industry

Regardless of whether or not the strategy described above is adopted by some steel producers in the United States, the industry's structure is likely to change drastically over the next ten to fifteen years. To some extent, future developments will merely represent the continuation of past trends, for example, in mini-mill expansion. Gradual quantitative changes eventually produce qualitative shifts, however. The U.S. industry is now drawing the full implications of trends that were already evident in the early 1960s; the rapid metamorphosis of U.S. Steel indicates how profoundly the industry's structure is now changing.

Forecasting the eventual outcome of these changes is difficult, since political factors (especially in the trade area) and managerial strategies are so difficult to predict. In spite of this, the arguments presented in the preceding section have been drawn together to describe an optimal structure for the industry in 1985, 1990, and 2000. The results are presented in Table 10–1, which also describes achievable performance improvements. If the strategy described in the preceding section were adopted by the industry's integrated sector, the structural result would approximate the data presented in Table 10–1. Regardless of the strategy pursued, however, substantial retrenchment is inevitable.

One of the chief uncertainties in the construction of such a forecast involves the import share. Imports are expected to take 22.5 percent of the U.S. market in 1985 and 25 percent in 1990 and 2000, while the source of imported steel slowly shifts to developing countries. If integrated steelmakers are able to boost their efficiency significantly, this development plus the increased share of the already efficient mini-mill sector could restrain the growth in import share after 1990. Barring adequate cost reductions on the part of integrated producers, however, only increasingly severe trade barriers will limit imports to the levels described here—at great cost to American steel consumers. Under different assumptions (integrated producers continue to lose ground competitively and trade barriers are eliminated), imported steel could claim up to 40 percent of the U.S. market by the turn of the century.

Within the United States, the most gratifying trend in the 1980s and 1990s will be the increased strength of the mini-mill sector. The dynamism of its nonintegrated sector is one of the strongest potential advantages of the U.S. steel industry. In 1981, mini-mills accounted for about 15 percent of total U.S. steel shipments; by 2000, they should be competitive in up to 40 percent of the U.S. steel market. Barring spectacular increases in scrap prices (an unlikely eventuality), mini-mills are projected to capture 25.4 percent of domestic shipments by 1990 and 34.6 percent by 2000. Mini-mill expansion will be accompanied by increased electric furnace production by integrated companies, putting pressure on scrap supplies and encouraging the development of alternative new technologies such as coal-based direct reduction and plasma steelmaking.

The continued onslaught of the mini-mill sector will be accompanied by further structural changes. Mini-mills will be active in a far wider range of products. The larger minis will exceed one million tons annual capacity—a mark already passed by Chapparral Steel—and will be more vertically integrated, controlling captive sources of scrap or directly reduced iron. Together with increased electric furnace production by integrated producers, these trends will blur the distinction between the two sectors. The larger and more successful mini-mills will increasingly take on the specialized character of market mills, gaining substantial economies of scale and serving widely dispersed consumers of well-defined product lines. This will dramatically increase their efficiency and labor productivity; at present, labor productivity at the best market mills is 30 percent better than average mini-mill practice. Labor requirements of less than one manhour per ton will be common in the best mini-mills by the end of this century. On the other hand, the increased size and complexity of mini-mill operations will also increase their vulnerability to market downturns, further blurring the distinction between the two sectors.

The obvious implication of these developments for integrated producers is a decreased share of the domestic market. The effects of this shift on integrated capacity will be aggravated by the fact that integrated performance must improve drastically if those firms are to compete. Increased use of continuous casting (nearly 100% by the year 2000) and other measures should improve yields from 0.73 in 1980 to 0.87 in 2000. Increased yield, coupled with declining market share and the commitment to higher operating rates, portends a drastic decrease in the raw steel capacity requirements of U.S. integrated producers.

The full implications of these trends are distilled in Table 10–1. Integrated steel shipments will decline steadily, from 74 million tons in 1980 to 66.5 in 1985, 62 in 1990, and 59.1 in 2000. Assuming a steady improvement in average operating rate, from 85 percent in 1985 to 90 percent in 2000,[8] and the yields listed in Table 10–1, this implies that integrated steel capacity should fall from the late 1982 level of 130 million tons to 105.5 in 1985, 91.1 in 1990, and 75.5 in 2000—a total decline of almost 40 million tons by 1990 and almost 55 million tons by 2000. Mini-mill expansion will partially offset this decline, but the reduction in total U.S. raw steel capacity should still be substantial.

Table 10–1 also describes necessary performance improvements in labor productivity. As the data indicate, productivity is assumed to increase at an annual rate of 4 percent in mini-mills and 3.5 percent in integrated plants. These results reflect product shifts from the integrated to the mini-mill sector, enhancing overall growth in integrated productivity while worsening the performance of the mini-mill sector.

All of these trends (productivity improvements, slow market growth, high import shares, improved yields, etc.) imply sizable decreases in steel industry employment, from 396,000 in 1980 to 185,000 in 2000—a decline of more than 50 percent in twenty years. This rather stark result, however, actually represents merely the exacerbation of existing trends.

The projections presented here do not represent maximum likelihood estimates of future steel industry performance. Rather, they represent the authors' assessment of the changes required if the American steel industry is to restore its profitability and reverse its secular loss of competitiveness. Integrated steelmakers may resist the future described here, but there appear to be few alternatives—none that the market will support. Government subsidies and increased protection from imports could slow the pace of capacity reductions, productivity improvement, and employment losses; but such steps would merely postpone and worsen the inevitable retrenchment.

POLICIES FOR THE TRANSITION

The inadequacy of the industry's strategic response to the market conditions that are forcing contraction has been matched by the bankruptcy of the policy environment. As is the case with steel firms, the government has sought to cope with the short-run symptoms of decline. Both parties have resisted recognizing the underlying dynamic within the industry, so that there is no shared, constructive perception of the goals that policy should seek to achieve. De facto, the U.S. government has accepted the eventual contraction of integrated steelmaking in this country. Yet this inertial tendency has been confused and contradicted by a long series of ad hoc interventions to mollify the political constituencies that resist retrenchment. Specific steel policies, generally the response to exhausting lobbying campaigns, have camouflaged rather than addressed the underlying tendency of the market.

We will not attempt to describe a specific set of policies that would together constitute a coherent steel policy. In our view, the chief task facing policymakers in regard to steel—and other industries as well—is the establishment of a positive and progressive environment in which specific programs can bear fruit. In the past, steel policies lacked an overall vision of the industry's future. Instead, the government has been committed to the policy-strategy consensus that sought to defend the status quo—in sharp contrast to the dynamic consensus that characterized the Japanese and Canadian steel sectors from 1955 to 1975.

Both the government and the industry must accept that profound structural changes within the American steel industry are now inevitable. Even with the best of intentions, the option of maintaining the American steel industry in its present form is a luxury that is unavailable to either the industry or the government. Policies designed to maintain the status quo have little prospect of success and would impose a heavy penalty on the economy as a whole, particularly steel-consuming sectors. Retrenchment, rather than preservation, is the task that now confronts the integrated sector of the U.S. steel industry; this must become the basis for the industry-government consensus. At a minimum, the government's role should be to resist policy initiatives that delay major changes within the industry; more actively, the government could seek to accelerate the restructuring process.

Plausible arguments could be and have been raised to justify governmental intervention to protect the industry from decline. First, it has been argued that governmental policies have greatly contributed to the industry's loss of competitiveness and that a neutral or compensatory policy environment could reverse this process. Second, the importance of the industry to the national economy (or to national defense) has been cited to justify protectionist measures. Finally, it has been argued that the government has a responsibility to stem the loss of jobs that is the indirect product of the industry's contraction.

None of these considerations justify a governmental effort to retard the restructuring process. It is true that governmental actions have been one element in the milieu that has characterized the steel industry's decline: the tax system has been biased against capital-intensive industries, particularly in an inflationary environment; incomes policies have disrupted wage and price formation, particularly in steel; an overvalued dollar in the fixed exchange rate era encouraged import substitution; and so on. Yet it is doubtful that different policies could have radically altered even the rate at which steel has lost its importance for the U.S. economy. One must look elsewhere to explain this decline, as this book has shown. Nevertheless, the sectoral impact of government policies should be clarified. An explicit governmental commitment to this type of research program could increase policymakers' awareness of the heretofore unintended consequences of policy actions.

Steel's importance to the national economy (and to national defense) is easily dismissed as a policy issue. Even assuming significant retrenchment in the integrated sector, domestic output will continue to outweigh imports in the U.S. market (see Table 10-1, for instance). Furthermore, steel is available from a diverse enough set of countries that there is no real danger of an "OSEC" controlling the American steel market. Defense industries purchase a small share of steel output, and shipments could be diverted from other uses in a national emergency.[9] Our assessment of developments within the domestic steel industry does not support governmental intervention for defense purposes.

Probably the most persuasive argument for governmental intervention in defense of integrated steel production concerns the government's responsibility to the industry's labor force. This will be a critical social question for the foreseeable future, since several of the nation's traditional industries face retrenchment as the United States becomes more integrated into the global economy. Nevertheless, structural unemployment among previously skilled and well-paid workers is a problem that extends beyond individual industries, so that the response should not be industry specific. As far as steel is concerned, the problem will exist regardless of the steel policy favored by the federal government: either the industry will continue to perform poorly, with bankruptcies and facility closures, or it will restructure radically (boosting productivity) in order to regain international competitiveness. In either case, the labor force will shrink.

Rising unemployment among steelworkers cannot be solved by measures designed to retain jobs in the industry. In Europe, governmental intervention for this purpose has led to uniformly disastrous results: inefficient operations were artificially maintained, perpetuating excess capacity and poor profitability; unprofitable operations became addicted to subsidies, so that their demands on public treasuries continually increased; and healthy operations were eventually undermined to the point where they too need governmental support to survive. For example, European subsidies to integrated companies helped push Korfstahl, a German firm that has been a world leader in mini-mill technology, into bankruptcy in 1983. This sort of policy environment, which strangles the firms and technologies that are emerging to replace outmoded and uncompetitive operations, imposes an enormous penalty on society as a whole. Moreover, government intervention to save disappearing jobs in the integrated sector only postpones rather than avoids the eventual crunch; the British Steel Corporation has provided a good example of this—at great cost to the British treasury.

This does not mean that reliance on the market is an adequate response to steel industry unemployment. Rather, our argument is that the ultimate solution to the problem will not be found in the steel industry. The policy problem revolves around the extent to which government is prepared to retrain displaced workers and either attract new industries to steel-producing regions or encourage workers to relocate. Past federal programs designed to deal with structural unemployment have been disappointing. Nevertheless, ways must be found to improve such programs, given the present magnitude of this problem.

While retraining is critical, it is also crucial to address the fact that for most displaced steelworkers, employment in the so-called growth industries will entail an enormous reduction in income; the steelworker who moves to electronics assembly, for instance, would probably see his or her hourly compensation cut by a staggering two-thirds. To a great extent, this is the gloomy reality behind the high-tech solution to the problem of basic industries. This is the most critical political issue produced by the decline of the American steel industry and one of the principal reasons there is a strong political constituency for protection of the steel status quo. This makes it all the more crucial that policymakers confront the issue, however politically unpopular, in order to establish an environment in which the necessary structural changes are encouraged. In the long run, the economic growth and increased efficiencies associated with such changes are the best guarantor of rising incomes. Evidence of this can even be found in the steel industry: according to Nucor's chief executive, the average income of that company's employees exceeded the industry average in 1981.[10]

If we reject the arguments in favor of defensive government policies (and thus accept the need to accelerate the adjustment process), this still fails to establish an environment that could nurture a progressive strategy-policy consensus. Even without the political pressures that resist change, the U.S. government is institutionally ill equipped to adjust its policies according to changing conditions. The government is no more apt than the industry in coping with structural changes.

Legalistic inflexibility is the principal symptom of the government's rigidity—akin to the industry's adherence to the massive modernization strategy. Antitrust rules, for instance, are applied as though integrated steel firms still dominated an essentially autarchic market.

Since restructuring and contraction are now clearly on the industry's agenda, antitrust restraints ensure that the process will be more protracted and irrational than would otherwise be necessary. If firms were allowed to coordinate the restructuring process, the prospects for an optimal result could be greatly increased—assuming that the industry truly commits itself to contraction and that adequate competitive pressures are maintained by imports and by domestic mini-mills. Without such coordination, individual firms will be reluctant to risk the losses in market share that would result from their own retrenchment if competitors failed to follow suit. Furthermore, coordination could ensure that the best facilities are saved, insofar as this is feasible on an industrywide level.

The specific features of a waiver of antitrust rules—and the extent to which the government should play an activist role in this process—are less important than the fact that this is the level at which the discussion of steel policy should take place. Without a government-industry consensus, governmental offers to relax antitrust restraints (as in the Solomon Plan of the Carter administration) are announced in a vacuum, making no sound. Other regulatory policies exhibit similar rigidity, thus limiting the prospects for a coherent and far-sighted policy mix as well as ensuring that there is little basis for the realization of an overall consensus.

With or without a progressive government-industry consensus, however, policies should be judged according to whether they facilitate structural change within the industry. Tax policies, for instance, should be devised to encourage entry and to reward those integrated firms that rationalize. Insofar as the tax system's capital recovery provisions have penalized capital-intensive industries with long-lived assets (such as steel), the tax reforms of the Reagan administration (especially accelerated cost recovery) represent a significant step toward sectoral neutrality in terms of the tax burden. These should help the industry. Tax "reforms" that block structural changes are not helpful, however. Generous depreciation allowances, especially when combined with safe-harbor leasing (which allows the sale of tax credits by unprofitable firms), retard structural changes by encouraging existing firms to hang on. Safe-harbor leasing is particularly damaging to the steel industry, since it bolsters firms' already excessive commitment to the massive-modernization strategy. These features of the tax system tend to reinforce the status quo; this is compounded by high marginal tax rates (offset to some extent by the promise of eventual capital recovery), which make it more difficult for new firms to enter the industry. Yet it is entry by nonintegrated producers that, more than anything else, is transforming steel production from a drain on the economy into a revitalized asset.

Other elements of the tax laws should be evaluated in terms of whether or not they foster increased competition and innovation. Raw materials depletion allowances should be reexamined in view of the extent to which they artificially encourage vertical integration and discourage the market pricing of inputs. Market pricing is crucial to improved resource allocation and thus to improved efficiency—a point that is particularly relevant to U.S. patterns of iron ore consumption. While American steel firms are more or less stuck with expensive domestic ore sources at the present time, the tax system should encourage the eventual disengagement of iron mining and steel production as a means of improving efficiency in both sectors.

Competition speeds the process of industrial adjustment, and that is one of the principal reasons the government should encourage mini-mill entry. The same line of reasoning should be applied to steel imports, in spite of the political attractiveness of trade barriers. The U.S. government should vigorously counteract the subversion of the market mechanism in international steel trade. Massive governmental subsidies, particularly in Europe, have perpetuated the world steel crisis and have intensified import pressure on the U.S. market. Targeted measures, for example, countervailing duties on subsidized imports, are the best means of protecting private domestic producers from the effects of such market distortions. Unfortunately, this sort of targeted intervention has not occurred. Instead, steel trade policy has drifted toward the integration of the U.S. market into a global steel cartel. Extensive quotas on steel imports to the United States will not "save" the domestic industry; rather, they will prolong and exacerbate the adjustment process, at great cost to the U.S. economy. The competitive spur provided by imports is a significant inducement for the structural changes that are needed; in fact, steel imports have probably been the principal factor dismantling the American steel oligopoly. This suggests that imports are a threat only if the goal of the industry-government consensus is defined in terms of the status quo.

Defense of the status quo now implies defense of poor profitability, lagging competitiveness, and persistent decline. Thus, there is no justification for the defensive posture of either the government or the industry; the prospect of significant changes should attract both parties, so that the objective basis for a progressive consensus exists. Mutual hostility must be overcome, uncertainty about governmental policies must be eliminated, and the time horizon of policy commitments must be lengthened. Most importantly, lobbying campaigns must cease to be an attractive substitute for efforts to boost competitiveness in the marketplace. If such an environment can be achieved, both the industry and the government will be less likely to concentrate on ad hoc measures designed to preserve the status quo.

Given a progressive industry-government consensus, the industry may be able to restructure and to adopt more appropriate technologies—an area in which governmental support for R&D would be highly beneficial. This represents the industry's best hope for reversing its competitive decline. In fact, within the context of the global steel market, the competitive struggle now revolves around which countries will be able to forge a policy-strategy consensus that encourages restructuring. The structure of the American steel industry has changed drastically since the 1950s, so that the adjustment process is already fairly well advanced in this country. While retrenchment threatens individual firms and involves difficult reductions in employment, the prognosis for the industry need not be a gloomy one. All parties should be able to agree on the goal: a profitable, competitive industry. The means to this end are more problematic, but even here there is enough evidence to define the strategies and policies that will position the industry to anticipate its market rather than lag behind.

CONCLUSION

The chief conclusions of this study can be summarized as follows:

1. The world steel industry is presently undergoing major structural changes, characterized by slow market growth, locational shifts in production and consumption, and altered managerial and technical requirements. These developments began as early as the 1950s in the United States and in the late 1960s in other developed countries. Structural changes have accelerated in the 1970s and 1980s.

2. American integrated steel producers have lost ground, first to foreign producers and then to new domestic competitors (the mini-mills). The mini-mill phenomenon represents a highly significant market response to the altered conditions of the American steel industry. As a result of these new entrants, the oligopolistic structure of the American industry has collapsed—a process that has produced a steady decline in the market share, profitability, and employment of integrated firms.

3. The failure of American integrated producers can be partly attributed to relatively high labor and raw materials costs. However, the major problem has been inadequate performance—especially in terms of labor productivity and the efficient use of limited capital resources.

4. A careful evaluation of the sources and causes of comparative performance (in Japan and the U.S., for both mini-mills and integrated plants) reveals that superior performance has been obtained, and is to

be sought, in the rapid adoption of superior operating techniques, major new technologies, high operating rates, and, to a lesser extent, scale improvements.

5. The inadequacy of U.S. integrated performance can ultimately be traced to the lack of a well conceived and coordinated strategy. This failure has been associated with a persistently optimistic view of market prospects and with the traditional price maker behavior of an oligopoly. The steel oligopoly has tended to emphasize price increases–which frequently depended on the industry's ability to affect government policies–rather than cost reduction. Given this background, competition within the steel industry emphasized struggles over market share, so that producers sought to maintain sufficient capacity for boom markets and to produce a full range of products. Performance improvements were sought via a strategy of massive modernization, that is, the large-scale, capital-intensive replacement of all facilities. This strategy has been forestalled by limited funds.

6. The integrated strategy has perpetuated excess capacity and the halting adoption of new technologies. This makes it extremely difficult to achieve real cost reductions and thus ensures a persistent decline in competitiveness vis-a-vis other successful integrated industries (e.g., the Japanese) and mini-mills.

7. Of the world's major steel producers, only Japan has successfully pursued the strategy of massive modernization and building ahead of the market. In the 1950s and 1960s, Japanese success was achieved because of very rapid market growth and the continuous application of new technologies. The Japanese strategy is inappropriate in a slow-growth market. An alternative strategy is essential under such conditions: downplay size, concentrate on a well-defined market, seek high operating rates, and obtain scale economies through specialization. This type of strategy has been pursued by both the Canadian steel industry and U.S. mini-mills; the latter have refined this strategy to emphasize the rapid adoption of capital-saving technology. This strategy holds the key to improved steel performance in the 1980s and 1990s.

8. The problems of U.S. integrated producers have been complicated by a negative policy environment, in which the government-industry consensus has been based on a reactionary commitment to the status quo. Under these circumstances, policy changes have been awarded to intense lobbying efforts. These "victories" have done little to improve the industry's long-run performance and have often cost dearly in terms of the price paid for political support. In fact, their principal result has been to expose the bankruptcy of the present policy environment. Other countries have established more progressive policy milieus, with greater

recognition of the need for shared goals and consistent policies and strategies.

9. The structural changes now underway portend continued slow market growth, increasingly sophisticated and diverse product requirements by steel consumers, increased import penetration, greater market capture by mini-mills, and substantial decreases in employment. Under these circumstances, integrated firms' strategies must be completely revamped. More appropriate strategies would seek operating improvements and reduced capital costs. The means to these ends would be substantial capacity reduction (boosting operating rates even in weak markets), product specialization, improved work practices, and the rapid development and implementation of superior operating techniques—especially improvements that are capital saving. Reduced input prices are a prerequisite for renewed profitability, but improved performance is the long-run solution.

10. Specific policy changes are less important than the forging of a constructive government-industry consensus built around accelerating structural changes rather than maintaining the status quo. The job losses associated with these structural changes represent a significant policy issue, but they cannot be solved by measures within the steel industry. Antitrust rules, tax regulations, import restrictions, and so on should be evaluated on the basis of whether they accelerate rather than retard change. Nevertheless, it is essential that the industry forsake the view that its salvation lies on Capitol Hill rather than in the plants.

Collectively, these points render a severe judgment on the present structure and performance of the American steel industry. The inevitable recovery from the tremendously depressed levels of 1982 may encourage the view that this judgment is excessive, that minor modifications in industry strategy or government policy could revive the industry, its traditional structure intact. Such optimism, however, would only make the next downturn more severe. The decline of the integrated sector of the American steel industry has deep roots, a point that is unequivocally confirmed by a long-term assessment of its competitive performance. The watershed for the industry came around 1960, when the entry of imports and domestic mini-mills signaled the beginning of the end for the oligopoly that had dominated steel production in the United States since the beginning of this century. By 1982, U.S. Steel's diversification away from the industry confirmed that this process was complete, marking the end of one era and thus the beginning of another. Some integrated firms may pursue a strategy of leaving the industry. Those that seek to remain, however, must alter their strategies or face extinction.

Notes

1. Lewis Carroll, *Through the Looking Glass* (London: Oxford University Press, 1971), p. 145.
2. The treatment of this issue is inevitably somewhat awkward, since the story of the industry's price maker status is largely the story of the U.S. Steel Corporation's price leadership. There are several more progressive integrated firms whose performance is superior to the industry norm that is described here.
3. Frederic M. Scherer, *Industrial Market Structure and Economic Performance* (Chicago: Rand McNally, 1970), p. 169.
4. See American Iron and Steel Institute (AISI), *Steel at the Crossroads* (Washington, D.C.: AISI, 1980), p. 44; and "Report to the President by the Steel Tripartite Advisory Committee on the United States Steel Industry" (Washington, D.C.: U.S. Department of Commerce, September 25, 1980), p. 8.
5. Calculated from data presented in U.S. Department of Commerce, Bureau of Economic Analysis, "Survey of Current Business," vol. 62, no. 7, July 1982, p. 73.
6. More exotic steel technologies are described in Office of Technology Assessment, *Technology and Steel Industry Competitiveness* (Washington, D.C.: U.S. Government Printing Office, 1980), pp. 194-214.
7. Work rules are negotiated on a plant-by-plant basis, while compensation is determined at the national level. This makes it more difficult to establish the extent to which work rules are being altered by the industry's present crisis; the comments in this chapter may understate steel firms' efforts to effect changes in work rules and procedures.
8. Operating rates have traditionally been higher in mini-mills than in integrated plants. A differential of 4 percent is assumed for 1985, 2 percent for 1990, and zero for 2000.
9. See Robert Crandall, *The U.S. Steel Industry in Recurrent Crisis* (Washington, D.C.: The Brookings Institution, 1981), pp. 98–103.
10. "Mini-mills, Maxi-profits," *Time*, January 24, 1983, p. 59.

APPENDIX A
THE PRODUCTION PROCESSES OF THE STEEL INDUSTRY

Steel production involves the chemical alteration of raw materials in order to introduce the desirable physical properties of steel, a transformation that is effected in ovens and furnaces (the hot end). Once this is accomplished, the steel must be shaped into products that can be used by other industries: sheets, plates, rods, bars, and so on. This appendix describes the basic steps involved in steel production and explains some of the associated terminology that may be unfamiliar to readers. For a more detailed description of the steel production process and its myriad permutations, see American Iron and Steel Institute (AISI), *The Making of Steel* (Washington, D.C.: AISI, 1979).

Steel is a generic term for a group of iron products that have similar metallurgical properties. The production of steel involves the chemical alteration of iron ore, the oxidized form in which iron is found naturally. Iron ore must first be reduced (i.e., the oxygen removed) before it can be used; this is the goal of the ironmaking process. Even after it is reduced, however, iron has properties that limit its usefulness; in particular, the material tends to be brittle and to fracture unless it is carefully worked. The presence of impurities, especially carbon, are responsible for iron's limitations. The basic purpose of the steelmaking process is to reduce the carbon content in iron and thus to produce the more widely usable iron product, steel. By definition, steel is iron with a carbon content of less than one percent. This makes it both a much stronger and a much more malleable metal. Other desirable properties (weldability, corrosion resistance, ductility, etc.) can be achieved by the introduction of small amounts of alloying elements at the steelmaking stage. The presence of alloys distinguishes specialty (e.g., stainless) steels from commodity-grade (or carbon) steel. Increased control of metallurgical properties—through better understanding of the processes at work, through more responsive equipment, and through more sophisticated use of alloying agents—is the fundamental physical challenge facing steel producers.

The following brief discussion of steel production will treat each of its major steps in turn: raw materials preparation, ironmaking, steelmaking, the production of semifinished outputs, and the production of finished products.

Raw Materials Preparation

Before entering the ironmaking furnace, the various raw materials that make up the charge or burden that is introduced into the furnace must be prepared. Iron ore, the principal ferrous input, can be introduced directly into the blast furnace, but if the natural ore's ferrous content is low, it is generally upgraded. The ore may also be agglomerated to produce a superior blast furnace feed. One type of agglomeration is pelletizing, in which iron-bearing rock is ground down to an iron powder which is then mixed with clay and other materials and rolled into pellets, which can be transported easily. Pelletizing is normally a mine-site operation.

Agglomerates may also be produced at sintering plants, which are usually built at steel mills. Sintering plants transform very small particles of ferrous material, called ore fines, into sinter, which may also include limestone and coke breeze (dust). A sintering plant may also be used to recycle the substantial quantity of ore fines generated at a steel mill (e.g., flue dust). Whether sinter or pellets, agglomerates improve blast furnace operation because, as agglomerated particles, they permit a better flow of air.

The second major input into the blast furnace is relatively pure (metallurgical grade) coal, which must first be baked in coke ovens into coke, a charcoal-like substance suitable for use in blast furnaces. Unlike coal, coke burns evenly, does not clot as it burns down, and can support the burden inside the blast furnace without packing down and thus shutting off the air flow. A typical coke oven installation has about sixty individual ovens, which are roughly 40 feet long, 20 feet high, and only 1 to 2 feet wide. Inside the coke oven, metallurgical coal must be baked for about eighteen hours in the absence of air; the noxious gases produced by this process are captured and used as fuel or chemical inputs at other facilities. The coke is then pushed into a quenching car, where it is doused with water and then transported to the blast furnace area.

The third major component of the blast furnace burden is limestone, which serves as a fluxing agent, facilitating the elimination of impurities from the iron ore and the reduction process.

Ironmaking

The traditional means of making iron is in the blast furnace—a tower in which the burden is introduced at the top and through which molten iron flows as it is chemically reduced. Once it reaches the bottom of the furnace it can be tapped (i.e., drained) as molten iron. This tower is supported by a complex of other facilities, especially: the stockhouse, in which the burden is mixed; the conveyer system, which delivers the burden to the top of the furnace; and the stoves, which provide the blast of hot air that gives the blast furnace its name. When a furnace is in operation, a mix of ferrous material (chiefly pellets, ore, and/or sinter), coke, and fluxing agents is constantly and gradually introduced at the top of the furnace.

A blast of hot air (roughly eighteen hundred degrees Fahrenheit) is introduced lower down in the furnace and is thus blown up through the tower or stack. This hot air mixes with combustible gases, is collected at the top of the furnace, and is transported back to the stoves. The stoves, which look like farm silos, operate in cycles as heat transfer systems. The hot gases from the blast furnace are used to heat the network of bricks in a stove; once these are sufficiently hot, they are used to heat cold air from outside the system so that it can be introduced into the blast furnace. As the temperature of the bricks falls below the necessary threshold, the stove is reconnected to the hot gases from the blast furnace and is thus recharged. As a result, each blast furnace must be served by at least two stoves to ensure a constant flow of heated air (the blast).

Inside the stack, the hot air flowing up through the furnace heats the burden, beginning the combustion of coke and the reduction of the iron input. The heat increases as the materials drop through the furnace; at the point where air is injected from the stoves (through valvelike systems called tuyeres) the coke burns fiercely, reaching temperatures of 3000 degrees Fahrenheit and melting the now reduced iron, which drops to the bottom of the furnace. The furnace is being tapped constantly. Molten iron is allowed to flow out holes at the furnace base and to run down channels into specially designed railroad cars for transport to the steelmaking furnace, although some iron produced in the blast furnace may be sold as pig iron. Generally, only one hole is being tapped at a time in order to preserve a balance of materials throughout the stack; the holes are plugged by rapid injections of clay when tapping ceases. These plugs can be blown away when tapping begins again.

Recently, alternative technologies have been proposed or developed to circumvent the blast furnace by directly reducing iron ore (to produce directly reduced iron or DRI). Such techniques all involve the use of a fuel such as natural gas or nonmetallurgical coal; chemical reduction occurs at temperatures below the melting point of iron, so that the output of the direct reduction process is solid. DRI is a suitable input for electric furnaces and thus a potential substitute for scrap. Direct reduction techniques, although increasing in importance, are still only marginally cost effective except where the opportunity costs of natural gas are very low.

Steelmaking

Various techniques can be used to produce steel from iron. Three are commonly used in the United States: the open hearth furnace, the basic oxygen furnace, and the electric furnace. These different procedures can be compared not only according to the techniques used but also according to differences in input usage and differences in the time required for a complete cycle (from charging the furnace to tapping the molten steel), which is referred to as the heat time.

The open hearth was the dominant steelmaking technique in the United States through most of this century. Open hearth shops generally contain several furnaces, each served by a common network of support facilities. The open hearth can take a widely varying mix of scrap and molten iron, and these are mixed with fluxing agents (principally limestone) and alloying agents. These inputs are mixed in the open hearth, a shallow box that is coated with chemically treated bricks (refractories) designed to facilitate the required chemical reactions. The furnace is then heated by the combustion of natural gas, which melts the scrap and fuels the refinement process. By the late 1950s, heat times in open hearths were about twelve hours. Heat times have now been cut in half, largely through the use of oxygen lances similar to those used in the basic oxygen furnace (BOF).

The BOF—so called because the furnace is lined with basic refractories and uses oxygen—is based on a radically different approach. It is charged mainly with molten iron (hot metal), although its ability to accept scrap has risen slowly over time. Once the furnace is charged, an oxygen lance is dropped through the top of the furnace and blows a blast of oxygen on the molten material in the furnace. The presence of pure oxygen fosters a violent chemical reaction; this is adequate to maintain heat in the furnace, so that supplementary fuel sources are unnecessary. The reaction proceeds so quickly that heat times in a BOF have been cut to about an hour; the actual refinement of steel takes about twenty minutes. The rapidity of the process is the principal advantage of the BOF, although it also ensures that sophisticated metallurgical monitoring and control procedures are required. A now common refinement of the BOF is the Q-BOP, in which oxygen is injected through tuyeres into the bottom of the furnace.

Finally, the electric furnace is an alternative route to steel production—one that generally melts down steel scrap, although it can also melt a solid iron input like DRI. In an electric furnace, the scrap charge is introduced and is melted by electrodes introduced through the cover of the furnace. The heat time is determined by the size of the furnace and the voltage used. Electric furnaces were traditionally used for the production of specialty steels, but since the late 1950s they have become a highly competitive source of carbon steel as well. Heat times in electric furnaces have been reduced from about six hours in the 1960s to one hour in the most modern furnaces of the early 1980s.

From Steelmaking to Semifinished Inputs

Once steel has been produced in the furnace, it is tapped into a ladle. Various chemical transformations can be executed while the steel is in the ladle—an approach that is becoming increasingly common. Traditionally, the ladle is then carried to a teeming aisle, where a train of ingot molds is waiting. The ladle is then emptied (from the bottom, using a plunger) into the ingot molds. The train is then moved away from the steelmaking shop, and the molten steel is allowed to cool into ingots, which typically weigh between 15 and 20 tons. Once the ingots have solidified on the outside, the molds are stripped away from them. The ingots can then be stored until needed. Even if used immediately, they must be placed in a soaking pit, where they are soaked in heat so that their temperature is uniform throughout. The process of rolling ingots into finished products can then begin. Ingots may be used directly in forging huge products like generator shafts or may be rolled directly into heavy products like large structural beams. Normally, however, ingots are first rolled into semifinished shapes (blooms, slabs, or billets) in a process called primary rolling. These semifinished shapes can be distinguished by their cross-sectional dimensions: slabs have a rectangular cross-section (e.g., 5″ x 45″), while both blooms and billets have square cross-sections. The edge of a billet measures roughly five inches or less, while blooms are larger (up to 12″ x 12″). Semifinished shapes rolled from ingots often have defective surfaces which must be ground down or burned off (scarfing). In smaller shops, steel from the steelmaking furnace can be poured directly into bloom-shaped molds, bypassing the ingot stage.

Continuous casting circumvents these steps. With this technology, the ladle is emptied into a tundish, a holding vessel with several gates on its bottom side. The tundish is positioned above a set of molds, into which steel is released through the gates. The steel then pours slowly through the mold, solidifying as it goes, until it emerges as a solid, semifinished shape. The caster is traditionally curved, so that the steel which is poured in vertically leaves horizontally (although fully horizontal casters are now being developed). As the steel slowly leaves the caster it is cut into suitable lengths by means of a torch, a laser, or a shear. The semifinished shape can then be tested and conditioned (e.g., grinding) before it is rolled into a finished product.

From Semifinished Shape to Finished Product

The various semifinished shapes can be stored until needed, when they must be reheated and rolled into finished products. Energy savings are obtained by directly rolling semifinished shapes into finished products without allowing them to cool. The basic principles of the rolling process are the same for all products; semifinished shapes are squeezed and pulled through rollers, the form of the rollers determining the shape of the final product. Billets are rolled into simple products like bars, rods, and light structural shapes. Products generated by the first rolling stage can also be processed further. Hot-rolled bars can be cold finished to exacting tolerances and surfaces, wire rods can be drawn into wire, and so on.

Blooms can be rolled into billets or into heavier finished shapes—e.g., structural beams and rails. Seamless tubular products are made from tube rounds produced from blooms or billets. The tube rounds are heated and penetrated at their center by a plungerlike apparatus called a mandrel; the tube is then formed around the mandrel.

Slabs are the most significant semifinished shapes in terms of the finished products for which they are used. All flat-rolled products (sheets, strip products, and plates) are made from slabs. Plate mills are used to run slabs through a series of rolls, which reduce their thickness to the desired level. The procedures involved are similar to those used in hot strip mills, enormous structures in which slabs are rolled through roughly ten to fifteen rolls in order to produce the sheets eventually used to make cars, appliances, and so on. The first rolls turn slowly, squeezing and lengthening the slabs. By the time the slab leaves the hot strip mill it is traveling at a very high speed. At the end of the hot strip mill the hot-rolled sheet is rolled into heavy coils. Hot-rolled sheet can be sold directly to customers; normally, however, it is subject to further processing at the steel mill.

The first step in further processing is pickling and oiling, a process in which the hot-rolled sheet is run through an acidic bath to remove surface flaking (scale) and coated with oil to retard oxidation. The sheets (which are in the form of long, continuous coils) can then be run through cold mills. In the tandem mill, the hot-rolled sheets are squeezed through rolls to reduce their thickness (gauge) and to strengthen them. This process makes the sheets more brittle, so that they must undergo controlled heat treatment (annealing) to recover the malleability desired by customers. Batch annealing is the most commonly used process; coils are heated in the absence of oxygen for a specified amount of time–usually several days–in large containers or annealing furnaces. Continuous annealing, a much more rapid procedure requiring substantial capital expenditures, is widely used in Japan and can be found in a small number of U.S. plants. Finally, the cold-rolled sheet is run through a temper mill to provide it with desirable surface properties.

Cold-rolled sheet can also be coated to resist oxidation. Galvanizing coats the steel with zinc by running the coil through a zinc bath (one-sided galvanizing is a recent innovation), while the tin mill coats the sheets with tin, usually by an electrolytic process. Tinplate is used primarily by the canning industry. Other coated products can be produced by similar means.

Flat-rolled products (sheets and strip) are also used to produce welded pipe and tube. Narrow strips (skelp) are welded in a spiral pattern or longitudinally to form the tube, although the end product is not as strong as the seamless tubes produced from blooms.

APPENDIX B
IRON ORE

This study would be incomplete without some discussion of the American iron ore industry. The material presented below is based upon a major study of the world iron ore industry undertaken by one of the authors for the Canadian government and completed in 1982.[1]

From the 1870s to the present, U.S. iron ore production has been concentrated in the Mesabi area of Minnesota and the upper peninsula of Michigan. American iron ore reserves are generally jointly owned and operated by American steel and iron ore companies (with some Canadian representation). The steel companies "purchase" the final product, while the iron ore companies usually develop and manage the properties. Through the 1950s American steel companies relied almost exclusively on domestic iron ore, which was shipped from the mines (west and south of Lake Superior) to steel producers at the base of Lake Michigan and south of Lake Erie. By the 1950s, higher grade U.S. ores had been largely exhausted. Although substantial tonnages of low-grade taconite ores remained, this material could not be readily used with then known processes.

At that time the United States began to turn to foreign ore sources, and U.S. companies were in the forefront of ore discoveries worldwide. Also in the 1950s, American companies decided to embark upon major iron ore development in northern Canada (Quebec/Labrador), the nearest foreign source. This was an extremely costly undertaking, requiring railroad, port, and town construction as well as mine development in a remote wilderness area. The end product was a moderate grade ore, which was shipped down the St. Lawrence River and along the Atlantic coast to U.S. steel mills. Some Canadian ore was also shipped to Europe. By the mid-1960s over 20 percent of U.S. ore consumption came from Canadian sources.

Figure B–1. Trends in Iron Ore Production[a] (millions of long tons)

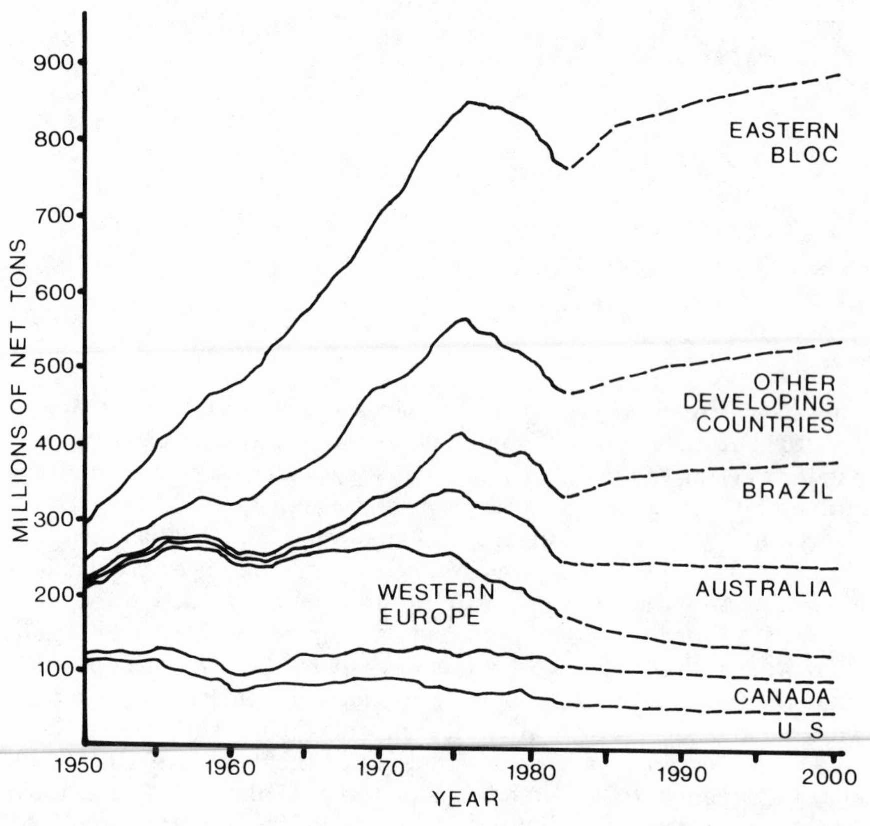

a. Five-year moving averages; trend projections.
b. Other developed countries; Australia, Canada, New Zealand, South Africa.
Sources: American Iron and Steel Institute (AISI), *Annual Statistical Report* (New York and Washington, D.C. AISI, various years); U.S. Bureau of Mines, *Minerals Yearbook* (Washington, D.C.: U.S. Government Printing Office, various years); U.K. Iron and Steel Statistics Bureau (ISSB), *International Steel Statistics*: *Summary Tables* (London: ISSB, various years).

Simultaneously, with the development of Canadian ore, new techniques were found to upgrade and process taconite ore in the United States; and major high grade ore discoveries were made in Brazil, Australia, and elsewhere. In the 1960s, with increased ore required to meet a stronger home steel market and to facilitate the shift from the open hearth to the basic oxygen furnace (BOF)—since the latter uses more ore and less scrap— the emphasis shifted to meeting expanded ore requirements by developing native taconite ores. However, imports did continue to increase through the mid-1960s.

The main foreign participants in iron ore projects outside North America were European and Japanese steel producers or trading companies. Iron ore production grew extremely rapidly in Brazil and Australia, and these areas now dominate world iron ore production (see Figure B–1). Offshore (i.e., non-North American) suppliers provided major tonnages to the United States in the 1960s and 1970s, when domestic or Canadian ores were inadequate, but were displaced by the latter sources in the late 1970s. Much Canadian ore was exported to Europe in the 1960s and 1970s, as domestic European ore production became increasingly uneconomic. More recently, however, Canadian ore has encountered heavy competition in the European market from Brazilian, Australian, and other ores.

Both U.S. and Canadian ores have a low natural grade, so that they must be upgraded and agglomerated (e.g., pelletized). Upgrading and agglomeration became prevalent in the 1950s and 1960s in order to produce an improved blast furnace feed and to compete with the much higher natural grade ores from Brazil, Australia, and other sources. Unfortunately for North American ore producers, equivalent grade offshore ores in Brazil and Australia have much lower costs of production than those in the United States and Canada because of the offshore ores' high natural grade, ease of extraction, and access to bulk transport.

Costs of production for comparative iron ore types and grades in the United States, Canada, Brazil, and Australia are shown in Table B–1. Table B–2 shows the cost of delivering these ore types to various markets. As the data show, f.o.b. production costs for equivalent grades and types of ore (pellets or sinter feed) are much higher in Canada and the United States than in Brazil and Australia. The U.S. disadvantage is so great that Brazilian ore, and to a lesser extent ore from other sources, can enter the Great Lakes market and compete with American, but not Canadian, ore on a full-cost basis. Brazilian ore could be imported via the St. Lawrence into the eastern Great Lakes and via the Mississippi into the western Great Lakes. The higher cost domestic ore imposes a significant cost burden on U.S. steel producers relative to their Japanese or European competitors. This burden results both from the central lo-

Table B–1. Comparative Representative Iron Ore Production Costs: 1980 (US$/long ton, f.o.b.).[a]

	U.S. (Minn.)	*Canada (PQ/Lab.)*	*Australia*	*Brazil*
A. *Sinter Feed (Approx. 62% Fe)*				
Labor	6.65	5.30	3.85	1.15
Energy	4.30	2.10	.75	.80
Raw materials	1.80	1.85	1.50	1.40
Other supplies & misc.	3.90	4.00	2.80	2.40
Transportation & handling (to vessel)	3.50	3.60	3.00	6.50
Operating costs	20.15	16.85	11.90	12.25
Depreciation & amortization	1.20	1.30	1.50	1.25
Interest	0.65	0.80	0.50	0.85
Royalties, management and local taxes	2.75	2.00	1.15	1.25
Total costs	24.75	20.90	15.05	15.60
Total revenue	27.00	19.25	17.50	18.00
B. *Pellets (Approx. 64% Fe)*				
Labor	9.15	7.30	6.25	1.90
Energy	6.25	3.15	4.00	4.10
Raw materials	2.75	2.85	2.60	2.75
Other supplies & misc.	5.90	6.15	5.00	5.25
Transportation & handling (to vessel)	3.50	3.60	3.00	6.50
Operating costs	27.55	23.05	20.85	20.55
Depreciation & amortization	1.80	1.95	2.25	2.25
Interest	0.95	1.25	0.75	0.75
Royalties, management and local taxes	4.20	2.75	1.55	1.85
Total costs	34.50	29.00	25.40	25.35
Total revenue	41.00	40.50	28.50	30.00

a. F.o.b. respective closest shipping ports: U.S. - Escanaba; Canada - Sept. Iles.; Australia - Port Hedland; Brazil - Tubarao.

Source: D.F. Barnett, *The Iron Ore Industry: Problems and Prospects* (Ottawa: Department of Energy, Mines, and Resources, 1982).

Table B–2. Comparative Delivered Iron Ore Costs at Selected Markets from Representative Sources: 1980 (US$/long ton, c.i.f.).

	U.S. (Minn.)	*Canada (PQ/Lab.)*	*Australia*	*Brazil*
A. *Sinter Feed (Approx. 62% Fe)*				
Cost f.o.b.	24.75	20.90	15.05	15.60
Delivered Rotterdam				
Transportation	—	7.00	10.00	8.80
Costs c.i.f.		27.90	25.05	24.40
Delivered U.S. northern habors				
Transport	—	3.25	9.50	8.00
Cost c.i.f.	—	24.15	24.55	23.00
Delivered Yokohama				
Transport	—	11.00	5.25	9.50
Cost c.i.f.	—	31.90	20.30	25.10
Delivered Lake Erie ports				
Transport [a]	6.90	5.75	17.35	15.75
Costs c.i.f.	31.65	26.65	32.40	31.35
Delivered Chicago				
Transport [b]	5.70	6.95	14.60	13.00
Costs c.i.f.	30.45	27.85	29.65	28.60
B. Pellets (Approx. 64% Fe)				
Cost f.o.b.	34.50	29.00	25.40	25.35
Delivered Rotterdam				
Transport	—	7.00	10.00	8.80
Costs c.i.f.	—	36.00	35.40	34.15
Delivered U.S. northern harbors				
Transport	—	3.25	9.50	8.00
Costs c.i.f.	—	32.25	34.90	33.35
Delivered Yokohama				
Transport	—	11.00	5.25	9.50
Costs c.i.f.	—	40.00	30.65	34.85
Delivered Lake Erie ports				
Transport [a]	6.90	5.75	17.35	15.75
Costs c.i.f.	42.40	34.75	41.75	41.10
Delivered Chicago				
Transport [b]	5.70	6.95	14.60	13.00
Costs c.i.f.	40.20	35.95	40.00	38.35

a. Via St. Lawrence River for Australian and Brazilian ore, including costs of reloading on to seaway vessels. Shipping via Mississippi River and reloading at New Orleans and Chicago would cost slightly more.
b. Via Mississippi River for Australian and Brazilian ore, including cost of reloading on to Mississippi barges for Chicago destination. Shipping via St. Lawrence River would cost substantially more (approximately $4 a ton).
Source: Same as for Table B–1.

cation of U.S. steel plants, which raises shipping costs, and from the high production costs of domestic ore. Coastal U.S. mills could benefit from cheaper offshore ores but would be more exposed to foreign competition in steel products.

Despite these cost relationships, U.S. companies have continued to use owned ores. The rationale for this is essentially as follows: delivered costs of owned ore can appear to be lower, especially if operating costs rather than full (or opportunity) costs are considered; depreciation and depletion tax benefits are available for owned ores; and security of supply is desired to avoid the prospect of shortages.[2] This rationale has become increasingly dubious. Iron ore is not a scarce material, and there are many high-grade sources available. Iron ore depletion allowances provide limited benefits if profits are low. There are (or should be) higher priorities and more profitable uses for scarce capital than mine development. Finally, offshore supplies are cheaper in some markets.

Were it not for vertical integration (ore through steel products), it is likely that offshore and Canadian suppliers would have a higher proportion of the U.S. market than the 10 percent and 20 percent, respectively, they had in the early 1980s. Vertical integration protects the domestic iron ore industry by short-circuiting the market determination of opportunity prices for ore and weakens the market signals on which the domestic steel industry bases its decisions. With a freer ore market, it is likely that there would be more imports (including Canadian) or that higher cost domestic ores would be eliminated from the market or both, so that the disparity between domestic and foreign production costs would be less.

Following steady growth in the 1960s after a decline in the preceding decade, U.S. ore consumption and domestic shipments reached peak levels in the early 1970s, as is shown in Figures B–2 and B–1. This is directly associated with the stagnant demand for steel in the 1970s, the slowdown in conversion to BOF steelmaking, and the increasing proportion of steel produced from scrap in electric furnaces. During the 1970s, offshore suppliers successfully increased their penetration of Canadian export markets in Europe. The net result has been excess North American iron ore capacity. North American iron ore producers (basically steel companies) initially responded to this development by reorganizing ore markets: Canadian ore was displaced on the Great Lakes by U.S. ore, and offshore ore was displaced on the East and Gulf coasts by Canadian ore. These steps were taken to achieve fuller utilization of owned Canadian and American facilities. In marked contrast to the trends in the rest of the world, U.S. steel producers turned away from offshore ore suppliers; the offshore (especially Brazilian and Australian) share of

Figure B–2. Trends in Iron Ore Consumption[a] (millions of long tons).

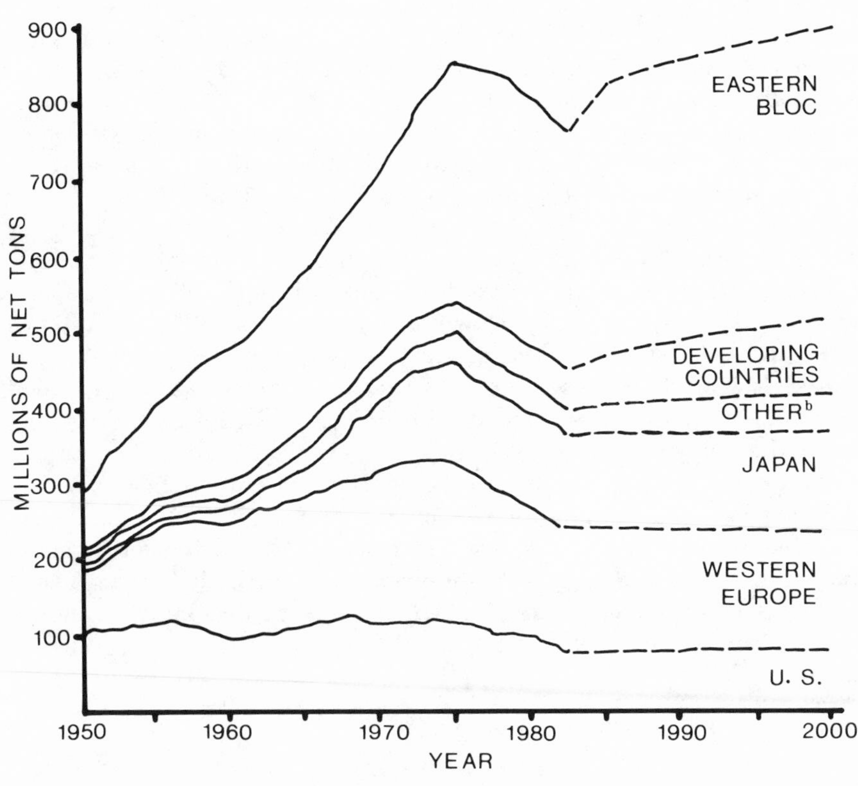

a. Five-year moving averages; trend projections.
Sources: Same as for Figure B–1.

U.S. ore consumption fell from 20 percent in the early 1970s to 10 percent by the end of the decade.

This "strategy" replaced lower cost ores with higher cost ores, largely because of the asset structure of U.S. integrated steel companies. As a temporary solution, this could be justified on the basis that domestic operating costs were lower than foreign full costs (i.e., prices). However, by the early 1980s came the inevitable realization that North American ore production could not be sustained, and the permanent closure of ore facilities began. The same factors are at work worldwide. Excess ore capacity accompanied the slowdown in the growth of world steel demand in the 1970s, and this is likely to be a problem throughout the 1980s—although the phasing out of European ore production and the natural depletion of ore bodies should eventually correct the imbalance, assuming no major new projects are funded.

Figure B–2 shows trends in world iron ore consumption by the major consumers. As the data indicate, consumption of ore in the United States grew little from the 1950s through the 1970s; over the same period, European ore consumption declined. Japanese ore consumption grew through the mid-1970s and then leveled off, as did that in the centrally planned economies. Ore consumption growth began later in developing countries, but only in developing countries has it continued to grow throughout the 1970s. As a result of slow projected growth in steel demand and, to some extent, increased use of scrap, global iron ore consumption is projected to grow slowly in the 1980s and 1990s, as is shown in Figure B–2. Iron ore consumption is projected to decline in the United States, while offshore imports are likely to increase their share of the market. This implies substantial decreases in U.S. iron ore production and employment, as shown in Table B–3. Significant and continued shifts in the location of world ore production are anticipated, as shown in Figure B–1. This is due to the superior grade and lower cost of ores available in Brazil, Australia, and other relatively new iron ore producers. Declining ownership of iron ore operations by North American steel producers is increasingly likely, given the cost burden imposed by domestic ores, especially on coastal plants.

Table B–3. U.S. Iron Ore Projections.[a]

	1980[b]	*1985*	*1990*	*2000*
Ore consumption (mmt)	94.6	82.0	76.0	73.0
Imports/apparent consumption (%)	29.0	30.0	32.0	35.0
Shipments to domestic buyers (mmt)	67.0	57.0	51.7	47.5
Domestic production (mmt)	72.2	61.0	55.2	50.5
Employment	18,000	14,000	11,900	9,300

a. Trend projections.
b. Five-year average centered on 1980.
Sources: Same as for Table B–1, with some updating. Iron ore projections based on steel consumption projections (Chapter 10) plus assumptions about share of scrap-based minimills and hot-metal ratios in integrated plants.

Notes

1. Donald F. Barnett, *The Iron Ore Industry: Problems and Prospects* (Ottawa: The Canadian Department of Energy, Mines, and Resources, 1982).
2. Full costs include operating costs plus depreciation, interest, royalties, and transport to the market. Continued production is justified in the short run provided operating costs are covered. Table B–1 shows that U.S. operating costs, plus transport to the market, have been lower than Brazilian full costs plus transport to the market. In the long run, however, there is no economic justification for continued preference for higher full cost domestic ores, although there may be some benefit in relying on secure domestic sources.

APPENDIX C
DATA METHODOLOGY

This appendix briefly outlines the data sources and basic methodology used in deriving the main data findings presented in this book, particularly the productivity analysis of Chapters 5 through 7 and the cost estimates in Chapters 2 through 7. It is not, however, a step-by-step guide to all the data transformations that were made. It will discuss only those procedures and results that would not be familiar to a specialized readership. In general, the data presented in this book were constructed on the basis of an extremely broad set of sources; a bibliographical list of the main published sources is presented at the end of this appendix.

All statistics of the type used in this book are estimates. Even the data that a firm generates on its own performance (some of which the authors, as consultants, have helped collect) are estimates. In the final analysis, even if individual components of our estimates may be questioned, we believe the analytical results will stand as the best possible reconstruction of past trends.

Labor Productivity by Process and Product

The base year, for which the most detailed productivity data were collected, was 1977. These data were used in constructing the trigger price mechanism (TPM) for the U.S. Treasury in 1978–79, a project in which D.F. Barnett was a major participant. Some pre-1977 data had been collected for a major study undertaken by D.F. Barnett for the Canadian government, and this data base was expanded in 1982–83 for this book. Chapter 5 describes in broad terms how the manhour per ton (MHPT) data by process were constructed using input-output (I-O) ratios to derive cumulative MHPT by product. In many cases, separate productivity data by product were available from the same sources, providing a check on the aggregation procedures used in constructing by-process results. Tables 5–2 through 5–4 show the results for only two products—cold-rolled sheet (CRS) and wire rods—in a single year (1980), although similar detailed relationships have been constructed for approximately twenty products for each year (roughly every third year for the United States) from 1958 to 1980. These additional data were used to calculate the results shown in Tables 5–5 through 5–10 and in subsequent chapters. The U.S. data apply to representative plants over the entire year, while the Japanese figures refer to average performance in October-November of each year. All MHPT calculations are at actual operating rates. Because of differences in data availability, different procedures were used for estimating productivity performance in Japan and the United States; they will therefore be discussed separately below.

Japan. The annual *Labor Productivity Statistics Survey* published by the Japanese Ministry of Labor contains virtually all the MHPT by process data (direct and total) used for Japan. The data are not published in the manner presented in this book, however; significant adjustments had to be made on the basis of the authors' detailed knowledge of steel processes. In fact, the data available from this source are more detailed than those presented here; many subprocesses are disaggregated (e.g., the BOF process is broken down into melting, charging, refining, etc. and finishing processes are broken down into crane operating, maintenance, etc.). These subprocesses had to be aggregated to obtain the process totals shown in Tables 5–2 through 5–4. For some processes, data were not available from this source (e.g., ore and coal handling and overhead) and had to be estimated. For recent years, estimates were made based on conversations with Japanese producers; estimates for earlier years were made by constructing indices linking more recent estimates with other process trends. For example, overhead was linked to overhead-worker ratios published in the Japan Iron and Steel Federation (JISF) monthly *Report of Iron and Steel Statistics*, while ore and coal handling were linked to subprocess material handling data for blast furnaces and coke ovens, respectively.

While the above sources provided data by process, no product aggregates were available. Aggregation was accomplished using the input-output relationships described in the tables. Most of these relationships are published in JISF, *Statistical Yearbook*, but some ratios (crude steel-semis; semis-hot rolled; hot rolled-cold rolled) had to be estimated. Crude steel-semis was estimated from data published in the *Labor Pro-*

ductivity Statistics Survey on the continuous casting share for each type of semifinished shape and from engineering estimates of yield losses for Japanese companies. The semis-hot rolled and hot rolled-cold rolled ratios were estimated from data on historical yields provided to the authors by Japanese steel companies (partially published in the *Statistical Yearbook*). These estimates concurred with other published estimates (e.g., Kawahito). More estimation was required for mini-mill than for integrated finished product yields, but the simplicity of mini-mill processes reduced the error potential to nominal levels.

The United States. In putting together detailed U.S. data, the major handicap was the lack of detailed published statistics from comprehensive sources. In order to make U.S.-Japanese comparisons by process, it was necessary to match U.S. MHPT by process data to the Japanese. Fortunately, the Japanese have adopted American engineering and process categories, so that both industries tend to adhere to the same process breakdown for labor statistics. Nevertheless, the compilation of U.S. data was like assembling a jigsaw puzzle, a very time-consuming process. Published MHPT data for the major processes (coke ovens, blast furnaces, etc.) are available from the *Census of Manufactures*, which is published irregularly by the U.S. Department of Commerce. Data for additional processes and years can be gleaned from the Department of Labor's Bureau of Labor Statistics (BLS). Most input-output relationships are published in American Iron and Steel Institute (AISI), *Annual Statistical Report.* Finished product yields are published by BLS and in consulting and engineering studies (e.g., Kawahito). Fortunately, from the perspective of data estimation, finished product yields in the U.S. industry have changed little between 1958 and 1980.

While these sources provide a great deal of data, there are many processes that they ignore. Some information on process MHPT has been published in engineering and consulting studies–e.g., A.D. Little, providing a much later but similarly comprehensive complement to the BLS data. Just as importantly, however, the authors relied on data amassed in years of private consulting (for steel companies and for the U.S. and Canadian governments). Because the U.S. data are less comprehensive than the Japanese, U.S. estimates generally describe representative performance while Japanese data describe average performance. U.S. data were superior to Japanese data, however, in regard to productivity by product. For the United States, product totals were available (e.g., from BLS) against which cumulative process data could be compared. This made it possible to fill in some process data gaps (e.g., in overhead MHPT).

Japanese data are available for each year (1958–80), but the limitations of U.S. data—particularly the irregular appearance of the Census reports—limited U.S. estimates to a smaller number of years (1958, 61, 64, 67, 70, 72, 74, 77, and 80). Data results for the United States have been submitted to selected companies for comparison with their own experience, and this procedure has confirmed the accuracy of the more recent (1970s) data. Greater uncertainty is attached to earlier U.S. process data, however. This is because U.S. published sources are incomplete, and it is not possible to confirm fully older process data since memories are short and past data lost. Nevertheless, the main process and product data are from published sources; only the direct MHPT and MHPT for some minor processes (e.g., subdivisions of hot rolling and cold finishing) are not fully substantiated. The direct-total MHPT breakdown is referred to at some points in this book, and these data must be regarded as subject to more error and treated with greater caution than process totals. Because of these possible shortcomings, this book has not concentrated on the implications of direct MHPT estimates or on other estimates that have not been confirmed by the cross-checking of various data sources.

The reader may question the degree of accuracy (second decimal place) of these data, and there would be some validity to this. The key published process data, however, are presented with such a degree of accuracy. For those statistics where estimation by the authors was required, second decimal places may be somewhat fanciful. However, since a combination of sources was used, we were faced with the alternative of either stating actual published data with less accuracy than was available or presenting estimated data with more accuracy than was fully substantiated. We chose the latter alternative.

Sources of Productivity Improvements and Advantages. Once the detailed process data had been assembled and aggregated for each product (as per Tables 5–2 through 5–4), it was possible to examine trends in productivity for major processes and selected products. This was done in Tables 5–5 and 5–6, which compare levels and changes in total MHPT for representative products and processes in selected years. The years used were chosen so as to minimize changes or differences in capacity utilization.

Although useful, Tables 5–5 and 5–6 do not enable one to allocate total productivity improvements among process sources. The process sources of productivity improvement were divided into either performance improvements in each major process or changes in process interrelationships (I-O ratios). Since both may be changing simultaneously, the allocation of productivity improvements to different sources becomes an index-number problem. I-O changes were split into two major groups: to the crude steel level (input use) and from crude steel to finished product (yields). Although the allocation of sources was complex in some cases, the general procedure was: 1) measure performance improvements for each major process weighted by the average I-O relationship in the periods compared; and 2) measure the effects of changes in I-O relationships between the two periods weighted by the average process productivity for the two periods.

The above procedure was used not only for allocating productivity improvements over time (Tables 5–7 and 5–9) but also for assessing Japanese integrated productivity advantages over U.S. integrated firms (Table 5–8) and mini-mill advantages over integrated firms (Table 5–10) at one point in time. For comparing Japanese to U.S. integrated productivity, differences in process performance are evaluated using average (Japan and U.S.) I-O relationships and MHPT as the respective weights. For example, performance superiority in continuous casting (CC) is the difference in CC MHPT multiplied by the average (Japan and U.S.) CC ratio, while "more CC" is the difference in CC ratios multiplied by the average MHPT. The effects of input use differences were estimated by multiplying differences in coke rates, hot metal ratios, and so on (to the crude steel ratio) by average coke MHPT, average hot metal MHPT, and so forth.

Capital Costs

Capital cost estimates are easier to obtain than are productivity estimates, and several of the sources listed at the end of this appendix provide capital cost data. Nevertheless, the capital cost estimates presented in this book are more detailed than those found in most other studies. For this we are grateful for the information on economies of scale in capital costs (and MHPT) provided by plant builders (Concast, Union Carbide, Blaw-Knox, Whiting, Ferrco, etc.) and by steel companies (Stelco, Rouge Steel, Florida Steel, etc.). Information provided by such firms was especially helpful in constructing historical estimates of capital costs (Figure 7–4 and Table 7–9). Historical costs were put in 1981 dollars by using the Department of Commerce GNP implicit price deflator for fixed nonresidential investment.

The new plant costs shown in Tables 7–3 and 7–4 are for state-of-the-art facilities built on existing plant sites (brownfield construction). Overhead and infrastructure were allocated pro rata to each process, with material handling allocated to facilities that use the material in question. Site preparation was excluded from the estimates.

For integrated CRS and mini-mill wire rods (WR), the estimates are based on separate plants of minimum optimal scale producing the products concerned, but for integrated WR it was assumed that WR was the marginal product in a plant of minimum optimal scale designed to produce products similar to wire rods (e.g., bars and light structurals). This was done because the market would not support an integrated plant of minimum optimal scale dedicated only to WR. This procedure may understate integrated capital costs for WR.

Comparative Production Costs

Throughout this book estimates of production costs, disaggregated into operating and capital components, were provided for various categories of steel products. Cost estimates were developed for new plants as well as for historical results, using the productivity estimates and capital cost estimates developed according to the methodologies described above. The method used to estimate total costs was developed by D.F. Barnett in constructing the TPM in 1977–78. Its early results were published in the Council on Wage and Price Stability's *Prices and Costs in the United States Steel Industry*. Besides the sources shown in the bibliographical list below, the cost estimates are based on data presented in company annual reports and submissions to regulatory agencies—e.g., the Securities and Exchange Commission in the U.S.—on numerous conversations and exchanges of data with steel companies, consulting firms, and plant builders, and on private consulting work.

The estimates of current and historical production costs are based on actual performance and accounting practices. The new plant costs represent best estimates of likely performance (in labor usage, energy efficiency, etc.) in state-of-the-art plants operating at a standard operating rate of 90 percent, unless otherwise specified. Standard amortization periods and standard debt-equity ratios are used in estimating financial costs. For estimating economies of scale in total costs and the cost effects of variations in capacity utilization (Figure 7–3), it was assumed that raw materials costs vary little with changes in scale and that labor costs are roughly 25 percent fixed and 75 percent variable. This division is based on regression relationships between capacity utilization and labor usage described in D.F. Barnett's *Economic Papers on the American Steel Industry*. Other operating costs were assumed to be 10 percent fixed.

The historic and current costs are representative of average costs (some of the costs were provided by a few representative plants). The new plant production costs do not, of course, represent actual practice, and such plants may never be built. Nevertheless, any reasonable range of alternative new plant cost estimates (e.g., Crandall) would give similar overall results.

List of Principal Sources

American Iron and Steel Institute (AISI). *Annual Statistical Report.* New York and Washington, D.C.: AISI, various years.

______. *Steel at the Crossroads.* Washington, D.C.: AISI, 1980.

Arthur D. Little, Inc. *Steel and the Environment: A Cost-Impact Analysis.* Cambridge, Mass.: Arthur D. Little, 1975.

______. *Environmental Policy for the 1980s: Impact on the American Steel Industry.* Cambridge, Mass.: Arthur D. Little, 1981.

Barnett, Donald F. *The Canadian Steel Industry in a Competitive World Environment.* Ottawa: Department of Energy, Mines, and Resources, 1977.

______. *Economic Papers on the American Steel Industry.* Washington, D.C.: American Iron and Steel Institute, 1981.

Council on Wage and Price Stability (COWPS). *Prices and Costs in the American Steel Industry.* Washington, D.C.: U.S. Government Printing Office, 1977.

Crandall, Robert. *The U.S. Steel Industry in Recurrent Crisis.* Washington, D.C.: The Brookings Institution, 1981.

Federal Trade Commission (FTC). *Staff Report on the United States Steel Industry and Its International Rivals.* Washington, D.C.: U.S. Government Printing Office, 1977.

Japan Iron and Steel Federation (JISF). *Report of Iron and Steel Statistics.* Tokyo: JISF, published monthly.

______. *Statistical Yearbook.* Tokyo: JISF, published annually.

Japanese Ministry of Labor (JML). *Labor Productivity Statistics Survey.* Tokyo: JML, published annually. (Original in Japanese.)

Kawahito, Kiyoshi. "Sources of the Difference in Steelmaking Yield Between Japan and the United States." Murfreesboro, Tenn.: Business and Economics Research Center, Middle Tennessee State University, July 1979.

Marcus, Peter J., et al. *World Steel Dynamics: Core Report J.* New York: Paine Webber Mitchell Hutchins, 1979.

______. *World Steel Dynamics: Core Report Q.* New York: Paine Webber Mitchell Hutchins, 1982.

______. "World Steel Dynamics: Steel Strategist," various issues. New York: Paine Webber Mitchell Hutchins, various dates.

Mathtech, Inc. *The Effects of New Source Pollution Control Requirements on Industrial Investment Decisions*. Washington, D.C.: Environmental Protection Agency, 1978.

Office of Technology Assessment. *Technology and Steel Industry Competitiveness*. Washington, D.C.: U.S. Government Printing Office, 1980.

Temple, Barker, and Sloan, Inc. *Analysis of the Economic Effects of Environmental Regulations on the Integrated Iron and Steel Industry*. Washington, D.C.: Environmental Protection Agency, 1977.

U.S. Department of Commerce, Bureau of the Census. *Census of Manufactures, Industry Series: Blast Furnaces, Steel Works, and Rolling and Finishing Mills*. Washington, D.C.: U.S. Government Printing Office, various years.

U.S. Department of Labor, Bureau of Labor Statistics. *An International Comparison of Unit Labor Costs in the Iron and Steel Industry, 1964: United States, France, Germany, United Kingdom*. Washington, D.C.: U.S. Government Printing Office, 1968 and updated regularly. Japan added subsequently.

U.S. Treasury Department. *TPM Price Manual*. 1978-81, updated regularly. (Unpublished.)

INDEX

ABOUT THE AUTHORS

Donald F. Barnett is currently an industrial economist in the Industry Division of the World Bank, on leave from the University of Windsor in Canada, where he is Associate Professor of Economics. He received his PhD. in 1968 from Queens University in Ontario. Other full-time appointments include positions as Senior Industry Advisor to the Canadian government and Chief Policy Advisor to the Council on Wage and Price Stability and the U.S. Treasury, where he was instrumental in setting up the steel trigger price mechanism. Most recently, he was Vice-president and Chief Economist at the American Iron and Steel Institute. He has extensive experience as a consultant to various government agencies and to private firms (e.g., the Ford Motor Company and various steel companies in the United States and Canada). He has written and edited numerous books and articles and has traveled widely as a lecturer, concentrating on the problems of basic industries like steel.

Louis Schorsch is currently an associate analyst at the Congressional Budget Office, active in the industrial policy area. His steel industry background includes experience as a millwright apprentice at a large integrated mill and as Senior Economist at the American Iron and Steel Institute. He has taught at George Mason University and American University, where he is now working on his dissertation.